The Making of the African Road

Africa-Europe Group for Interdisciplinary Studies

VOLUME 18

The titles published in this series are listed at *brill.com/agis*

The Making of the African Road

Edited by

Kurt Beck
Gabriel Klaeger
Michael Stasik

BRILL

LEIDEN | BOSTON

Library of Congress Cataloging-in-Publication Data

Names: Beck, Kurt, editor, author. | Klaeger, Gabriel, editor, author. | Stasik, Michael, editor, author.
Title: The making of the African road / edited by Kurt Beck, Gabriel Klaeger, Michael Stasik.
Other titles: African-Europe Group for Interdisciplinary Studies (Series) ; v. 18.
Description: Leiden ; Boston : Brill, 2017. | Series: Africa-Europe group for interdisciplinary studies ; v. 18 | Includes bibliographical references and index.
Identifiers: LCCN 2016054805 (print) | LCCN 2016056162 (ebook) | ISBN 9789004336742 (pbk. : alk. paper) | ISBN 9789004339040 (E-book)
Subjects: LCSH: Roads--Social aspects--Africa, Sub-Saharan. | Travel--Social aspects--Africa, Sub-Saharan. | Africa, Sub-Saharan--Description and travel.
Classification: LCC HE367.A45 M35 2017 (print) | LCC HE367.A45 (ebook) | DDC 388.10967--dc23
LC record available at https://lccn.loc.gov/2016054805

Typeface for the Latin, Greek, and Cyrillic scripts: "Brill". See and download: brill.com/brill-typeface.

ISSN 1574-6925
ISBN 978-90-04-33674-2 (paperback)
ISBN 978-90-04-33904-0 (e-book)

This book is printed on acid-free paper and produced in a sustainable manner.

Contents

Acknowledgments

This volume is the outcome of a research project on the African long distance road named 'Travel and Roadside Communities', which the editors, together with Rami Wadelnour, have carried out since 2011. Our colleagues Osman M. Osman, Matthias Gebauer né Fleischer, Valerie Hänsch, Francisco Kofi Olympio and Florens Eckert have at one time or another also participated in the project. The project was funded by a research grant of the Deutsche Forschungsgemeinschaft (BE 2310/6-1/2/3). We wish to acknowledge our gratitude for the generous funding without which the project would have been unthinkable.

The project was situated within the larger framework of the Priority Programme 'Adaptation and Creativity in Africa – Significations and Technologies in the Production of Order and Disorder' (SPP 1448). This programme, whose participants are based at Africanist research institutions throughout Germany, has throughout the years of its existence developed into a unique breeding ground for creative thinking about Africa. We owe a large debt of gratitude to our colleagues within the programme for lively debates and intellectual inspirations.

Our own ideas about the African road have been much influenced by the participants of our workshops in Thurnau Castle in June 2012, in Wallenfels in February 2013, in Lisbon in June 2013 and in Kloster Banz in 2015. Most contributors to this volume have participated in one or more of these workshops. Tatiana Argounova-Low's work on Siberia, Erik Bähre's on South Africa, Jamal J. Elias' on Pakistan, Sven Kirschlager's on Mexico, Elísio Macamo's on Mozambique, Rodolfo Soares' on Ethiopia, Osman M. Osman's on the Sudan, Christian Ungruhe's on Ghana and Samuel A. Ntewusu's historical work on the Gold Coast, Erdmute Alber's on Dahomey and Jan-Bart Gewald's work on the early roads in Zambia could not be included in this volume. Still we derived much inspiration from their contributions, particularly during the Thurnau Castle workshop. Financial assistance has been provided by the Bayreuth International Graduate School of African Studies and the Institute of African Studies at Bayreuth. We wish to express our gratitude for the generous intellectual inspiration and the equally generous financial assistance we have received.

We owe further gratitude to Matthias Gebauer for drawing the maps and to Sophie Mew for the copy editing of this volume. Finally, we are much obliged to Brill's anonymous reviewers for their expert advice.

Notes on Contributors

Kurt Beck

studied anthropology, sociology and Islamic studies. He currently holds the Chair of Anthropology at the University of Bayreuth, Germany, and is a member of the Institute of African Studies, Bayreuth. Email: kurt.beck@uni-bayreuth.de

Amiel Bize

is a PhD candidate in anthropology at Columbia University. Her current research focuses on roads and traffic accidents, time and informal economies. Email: amiel@whatfelt.org

Michael Bürge

is a PhD candidate at the Department of Social and Cultural Anthropology, University of Konstanz, and was a junior researcher in the DFG Priority Program 1448 'Adaptation and Creativity in Africa'. In his work in northern Sierra Leone he focuses on il/legal economies, insecurity and uncertainty and issues of ethics. Email: michael.buerge@uni-konstanz.de

Luca Ciabarri

is a researcher and assistant professor in Cultural Anthropology at the Università degli Studi of Milan, Italy. In 2007–08 he conducted field research in Somaliland on commercial routes in the frame of a research project at the Max Planck Institute for Social Anthropology, Halle/Saale, Germany. In 2011 he was Hunt Postdoctoral Fellow – Wenner Gren Foundation. Email: lucaciabarri@gmail.com

Gabriel Klaeger

currently works and teaches at the Institute of Social Anthropology, Goethe University Frankfurt. He received his PhD from SOAS (University of London) for his thesis entitled *Speed Matters: An Ethnography of a Ghanaian Highway, its Perils and Potentialities*. Email: g.klaeger@em.uni-frankfurt.de

Mark Lamont

follows African road building and Kenya's new infrastructural dispensation with interest. He continues research for a book, *Speed Governors: Road Safety and Infrastructural Power in Postcolonial Kenya*, while having published on this topic in *African Studies*, *Africa*, *Anthropology Today* and *Technology & Culture* (with Rebekah Lee). He is at present working on an ESRC-funded project on Cultural Rights under the 2010 Kenya Constitution at the Open University. Email: marklamont2016@gmail.com

Tilman Musch

is currently leading a DFG research project at the University of Bayreuth on spatial concepts among Tubu Teda in Eastern Niger. He received his PhD from the Institut national de langues et civilisations orientales (INALCO, Paris) for his work on the construction of space among Buryat pastoralists. His research focuses on nomadism, spatial anthropology, mobility and ethnozoology. Email: tilman.musch@uni-bayreuth.de

Michael Stasik

is a research assistant at the Chair of Anthropology, University of Bayreuth, Germany, and a PhD candidate at the Bayreuth International Graduate School of African Studies. His research is on urban cultures and economies in West Africa. Email: michael.stasik@uni-bayreuth.de

Rami Wadelnour

is a research assistant in the *Roadside and Travel Communities* project, Chair of Anthropology, University of Bayreuth, and a PhD candidate at the Bayreuth International Graduate School of African Studies. His research interests are the pursuance of order on Sudanese hinterland roads, travel knowledge and practices, skills, informal learning, mobile communities and legal pluralism. Email: rami.wadelnour@uni-bayreuth.de

List of Illustrations

Figures

Maps

CHAPTER 1

An Introduction to the African Road

Kurt Beck, Gabriel Klaeger and Michael Stasik

An Epistemic Moment

Infrastructure is 'frequently mundane to the point of boredom', as an ethnographer of infrastructure once remarked (Star 1999: 377). Infrastructure, being the substructure for mundane projects, is used as a matter of routine while focal attention is directed towards the projects themselves. Roads, sewage and electricity networks are the enabling structures located in the background of these contexts of action and they tend to be taken for granted, to the point of becoming invisible (except, of course, for plumbers, engineers or electricians). Such invisibility is a consequence of infrastructure's apparently solid ontological background status. In moments of breakdown, however, the routinised use of infrastructure is disturbed (Larkin 2013: 336), the project at hand becomes blocked, and cracks may even develop in that taken for granted reality – and this is an epistemic moment when infrastructure requires our attention. Another opportunity to become aware of infrastructure arises when it is new and when disputes about it are not yet settled in routine. As we show in this volume, roads in Africa give rise to such opportunities on a regular basis.

Africa is the continent with the lowest road density worldwide, and many of the existing roads indeed require full attention when using them. From this we would argue that using the road in Africa cannot be taken for granted in the same way as it can and usually is in the North Atlantic world. Furthermore, the expansion of the road network in Africa has never been faster than in the years of the African economic renaissance in the new millenium. From this, we argue that the road also expands into and affects people's lives and daily experiences on an unprecedented scale. Africa's road regime is rapidly changing, and as such, it opens up equally rapidly changing and new opportunities for engaging with the road practically as well as for thinking and discussing the road – an epistemic moment in Africa's road history.

For Europeans who have grown up in an age of mass automobility, driving may have become so natural that it is almost unconscious, taken for granted and thus boring – except for, perhaps, when speeding. It is therefore easy to forget that motor roads, together with the rules and conventions for using them, are quite literally cultural constructions which have led to a complete reconfiguration of the social order: 'Around a relatively simple mechanical

 | DOI 10.1163/9789004339040_002

entity, then, a whole new civilisation has been built', wrote the sociologist Nigel Thrift about the car and its infrastructure (2004: 46). Recent historical scholarship reminds us of the excitement created by different visions, discourses and differing cultural attitudes as well as by the diverging interests of professionals, politicians and users struggling to shape the modern motor road (Mauch and Zeller 2008; McCarthy 2007; Merriman 2007).

In large parts of the African continent, being a latecomer to automobility and certainly not in an age of saturated mass automobility, the road remains exciting. As the road regime is still under construction in more than one sense, it is open for diverging interpretations. Many of the struggles that took place in the United States in the early 20th century and in Europe in the mid-20th century regarding the definition, the design and the appropriate use of the motor road can be observed in Africa today – an epistemic moment indeed.

Appropriation of the Road

The state of the African road regimes, however, should not be regarded as a stage in the repetition of North Atlantic automobile history. In Africa, automobility and the motor road have not been developed autonomously like in the early North Atlantic experience, but instead draw from North Atlantic and, in some regions, Gulf states' models. But – and this should caution us against a simplistic diffusionist model – the fact that motor roads are North Atlantic transfers to Africa should not lead us to presume that the emerging socio-technological orders of automobility in Africa are merely copies of the North Atlantic model, as has been widely assumed (Volti 1996). Besides the argument that even North Atlantic automobilities can hardly be reduced to one single model, even pre-scholarly experience indicates that these imported technologies have been adapted and modified in their African environment in multiple ways, and that they have been invested with differing meanings. In Africa, imported road-related technologies and their significations are still in a state of 'interpretive flexibility', as Science and Technology Studies describe the state of societal attitude towards technology that has not yet settled in routine and that is still open to new interpretations (Bijker, Pinch and Hughes 1989). Contributions to this volume substantiate our view that beyond the surface of adapting to an overall model of North Atlantic socio-technological order lie large creative spaces for differing constructions of roads, transport and traffic.

The process described in the above is what we call appropriation. Appropriation should be envisaged here as a dialectical process of turning something that is alien, new and unfamiliar into one's own. We argue that this process provides challenges for those who appropriate; challenges which have transformative potentials for the appropriating social milieu. New visions, opportunities and practices then arise. We also argue, however, that the appropriated thing itself, being surrounded by new practices and visions as to what it is and could be used for, is reconstituted as different from its source. Appropriation is therefore always a dialectical process, much as Marx envisaged the appropriation of nature, as man by changing nature in the same process changes himself (Marx 1968: 192).

We consider appropriation as important within the context of the African road primarily due to the fact that it is not only the state and state regulations that make roads in Africa, but also its everyday users. As we will explore below, these road and roadside users and practitioners *make* the road through their respective practices and interactions, and thus create distinctive orders of the road. Unlike in the contemporary North Atlantic road regime, these orders go hand in hand with high degrees of local agency, negotiation and creative appropriation.

From our perspective, road making not only encompasses the physical construction of the road, but also its communicative construction. Beyond engineering, 'road works' require imaginative, narrative, cosmological and sensual efforts. Even with regards to physical road construction, we contend that the majority of African motor roads come into existence through popular efforts and are not solely constructed by engineering companies commissioned by the state.

Characteristics of the African Road

This volume, then, addresses the motor road in Africa, in particular the long-distance road, and its co-production realised by those who act and interact within the realm of the road and roadside. To speak generally about 'the' African road would be of quite a large order, given that African roads span the whole spectrum of possible manifestations. Across the continent, one can find smooth throughways complete with impressive flyovers and interchanges; important though unpaved long-distance roads that turn into swamps after the rains; and simple desert tracks that have never experienced any construction machinery or perpetual construction sites where works never seem to be completed. Or one can find the roadblock where the journey might find a sudden end. Indeed, the overriding fact regarding the African road is its diversity. Such diversity is evident in the contributions to this volume, particularly when comparing Bize's account of traffic jams with Musch's and Wadelnour's accounts of desert tracks, or again, with Beck's account of a new highway. We do, however, argue that four broad first-hand observations about the African road are appropriate, particularly in comparison with the North Atlantic road.

Firstly, long-distance road networks in the North Atlantic regions tend to be finely differentiated according to their function of channelling high speed traffic. They have been sanitised from undesirables, such as pedestrians and animals, hawkers, slow moving vehicles, vehicles in a state of disrepair and non-motorised traffic. In contrast, the African road is often multifunctional. Functional differentiation at road-level is an ongoing process where the outcomes of contests regarding the order of the road and its legitimate use remain open. An Indian parallel is offered in Edensor's work (2004) which contrasts British 'motorscapes' (with elaborate driving regulatory apparatus and sharp differentiations between road and off-road) with India's paucity of formal rules, its different 'street choreography' and its vast assortment of animal drawn and motor vehicles that compete for the road simultaneously. Furthermore, as well as for driving, the African road is exploited for a wide range of uses, for instance, as a mobile market (Klaeger 2012a; Stasik 2013) and as a space for the propagation of beliefs (Klaeger 2009).

Secondly, descriptions of North Atlantic road users are usually centred upon lone drivers and their cars. Such intricate is their connection that they are often depicted as a sort of hybrid creature, neither a machine nor a person, but an assembled social being: the 'driver-car' (Dant 2004: 74; Sheller 2004). We believe that there is an important element of truth in such an image – the car being the extension of the competent driver's body, which, with its signals of status and intent, becomes the social skin of the driver. Beyond this mechanically and electronically mediated interaction between cubicles of steel and glass, however, there seems to be no space for sociality between humans. We agree that, in principle, the North Atlantic use of the road is mainly autonomous and individual, and that immediate sociality is further circumscribed by the built environment and by traffic rules. Marc Augé has even coined the concept of 'non-places' to characterise faceless stretches of asphalt without history, sociality or emotional attachment (Augé 1992; see Merriman 2004 for a critique). Road travel in Africa, in contrast, is largely a collective and often crowded experience on lorries and pick-ups, in mini-buses, buses and *taxis-brousse*. On roads less travelled, the motor vehicle may even grow into a home for passengers and crew, like a ship on the ocean. The interior of these vehicles can then be regarded as a moving social space exhibiting its own structure as it moves through other social spaces. Rare accounts of this are Stoller's portrayal of riding a Songhay bush taxi (Stoller 1989: 69–83) and Wadelnour's description of desert travel in the Sudan (this volume). The latter conceives of such vehicular social spaces on the road as travelling communities and describes their social texture through sharing and struggling for space.

Thirdly, North Atlantic public transport – virtually the only form of collective travel – tends to be centrally and formally regulated by fixed routes and timetables. The African equivalent, rather, is open to informal and often improvised collective management by a wide range of participants. This gives rise to an equally wide range of negotiations and social interactions. In this respect, Stasik's account of the ways in which buses and their passengers are despatched from one of Accra's central bus stations is instructive (this volume) as well as Wadelnour's account of passengers' involvement in making the road journey. Both case studies open up a perspective on public transport as 'artisanal' and in a state of 'creative disorder' (Godard 2002) with significant opportunities for improvisation and negotiation (see also Godard and Teurnier 1992). This artisanal organisation of transport and travel in Africa thus differs significantly from what Schivelbusch (1986) describes for North Atlantic public transport as the 'industrialisation of time and space'.

Finally, North Atlantic long-distance roads have been sharply distinguished from off-road spaces, whereas African roads and roadsides offer a variety of opportunities for interaction between the travel communities and roadside communities. Lamont (in this volume) provides a detailed description of such an opportunity by recounting the collective efforts of travellers and roadside residents in dealing with a broken-down vehicle. Encounters between travellers and roadside residents may also be institutionalised, as Beck (2013) demonstrates with the truck stops along the Forty Days Road in western Sudan. Interactions between mobile roadside vendors and passing travellers, as observed by Klaeger (2012b) along the Accra-Kumasi road, appear to be entrepreneurial, entertaining and risky at the same time. These ethnographic studies show how people's interactions bridge the space between road and roadside such as by aligning supply and demand or divergent paces and rhythms.

These comparisons may appear oversubscribed in the attempt to construct ideal types. For instance, North Africa, with its exceptional position within the continent, and southern Africa, arguably due to its Apartheid legacy, deviate from this pattern to a considerable degree. (And the reader should be aware that neither North African nor southern African cases are represented in this volume which refers to the *African* road in its title.) Yet we still find these comparisons useful as heuristic devices. They imply that in order to arrive at a meaningful description of the order of African long-distance roads, attention should neither be based on the perspective of the road alone, nor on the assumption of the functionally differentiated highway or on the presupposition of the limited sociality of the solitary driver. Instead a perspective should be developed which (i) captures the road and the infringing roadside as conjoint, which (ii) is able to accommodate road use from a range of different motives beyond driving, and which (iii) is open for all sorts of actors and communities involved in everyday road making practices. Such a perspective promises to provide insights into the unique spaces of encounter, practices and sociality that emerge on and alongside the African road. We contend that such spaces as well as their specific road-to-roadside orders are worthy of our close attention as social scientists.

The State of the Road

Africa made an impressive start into a railway regime of mobility in the late 19th and early 20th century, beginning in southern Africa and later in the Sudan, western and eastern Africa. As in the case of road development in much later years, railways still bear the imprint of a colonial economy. Certain lines were constructed to provide infrastructure for the first industrialised wars at the turn of the 20th century, and the system was generally designed to serve the extraction of raw materials from productive hinterlands to coastal ports. Today the total size of the rail network is estimated at 70,000 km – with nearly a quarter having fallen into disuse by the early 2000s. Most of the networks are single track and – because of the colonial design – have low interconnectivity except for in southern Africa where density is also highest (Gwilliam 2011: 86).

Since the 1960s, Africa's railways have been losing ground against road transport. In most African countries, roads have come to dominate the transport sector, carrying 80 to 90 percent of passenger and freight traffic (Gwilliam 2011: 17). The design of postcolonial African roads resembles that of the colonial railway design. Generally, the road network provides for good connections between the capital, large cities and ports within a state but for low cross-border connectivity. The length of Sub-Saharan Africa's main and secondary road network today is estimated at just over one million km. Together with an unclassified network of close to half a million km and an urban road network of nearly 200,000 km, the total road network is estimated at 1,735,000 km. The length of rural roads – for most part unclassified and unrecorded – is particularly difficult to determine. The most reasonable guess is that Africa has a network of one million km of designated rural roads (either tertiary or unclassified) that come not under national but local control, together with a huge network of 'undesignated' rural roads and tracks which defies even very courageous estimates (Gwilliam 2011: 19–22), but which is certainly much larger than the designated network (Gwilliam et al. 2009: 10). The undesignated roads are called 'orphans' in World Bank terminology because they have no formal owner (Gwilliam et al. 2009: 21). These are the roads which have been constructed by means of overseas development aid or by local communities and have since received no maintenance, as well as the roads that simply have come into existence 'under the wheel' (Beck, Wadelnour, both this volume).

These figures may appear impressive, but comparison with other world regions and putting the network's length and quality in proportion to population or land area may be sobering. According to recent calculations of the Africa Infrastructure Country Diagnostic programme, Sub-Saharan Africa has a total road network of 3.40 km per 1000 inhabitants compared with a world average of 7.07 km. In terms of paved roads this amounts to 790 m per 1000 people, which is one fifth of the world average (Gwilliam 2011: 22). The spatial density of roads in Africa is also significantly lower than anywhere else in the world.[1] It is certainly not an exaggeration of facts to claim that people in Africa live at the threshold of a 'walking world' (Porter 2002).

Considering an African international airport, the usual gridlock of a central business district in an African capital or an arterial road leading to populated suburbs, the picture certainly looks different. The African road regime is characterised by huge inequalities. African roads are concentrated in urban areas and predominantly connect large towns and ports. There are also regional disparities with southern African countries having high network densities and equally high rates of paved roads, while low income countries, particularly in central Africa, have extremely low networks of paved roads.[2] Especially in low income countries, road deterioration often outpaces maintenance capacities. A World Bank report suggests that on average about half of Africa's main road network, which is under the responsibility of the central state, is in good condition and twenty percent in bad condition requiring rehabilitation. When it comes to rural classified roads, which usually are under the responsibility of local government, about half is in bad condition on average and only 25 percent are in good or fair condition respectively. The undesignated 'orphans' receive sporatic maintenance at best (Gwilliam et al. 2009: 41). African countries on average spend double the amount for road construction of what they spend for maintenance, whereas as a rule of thumb both sorts of expenditure should be

1 The world average spatial density of roads is calculated at 944 km of roads per 1,000 km^2, whereas Africa has 204 km of roads per 1,000 km^2 with only one quarter paved, making for a spatial density of roads which is six percent of North America's (Gwilliam 2011: 22).

2 The average density is estimated at 10.7 km of paved roads per 1,000 km^2 for African low income countries, and as low as 100 metres of rural road, paved or unpaved, per 1,000 km^2 in the Democratic Republic of Congo (Gwilliam 2011: 22).

roughly equal to prevent deterioration of roads after some years. The authors of the report conclude that 'as a result, a significant number of countries are in a vicious cycle of low maintenance budgets leading to network deterioration, but without adequate capital resources to clear their escalating rehabilitation backlock' (Gwilliam et al. 2009: 53). Recurrent claims concerning connectivity should therefore be received with reservation. To discuss Africa in the fashionable tropes of mobility and flow would be a misrepresentation. It might make sense in certain environments and milieus, yet for many other environments, a description in terms of severely limited mobility and 'roadlessness' is much more accurate.

As in the case of roads, estimates for motor vehicles in Africa are weak and also have to be considered with caution. Clearly, the number of motor vehicles in the last decades has increased significantly. In 1980 – prior to the generally impoverishing impacts of structural adjustment programmes throughout Africa – the total number of motor vehicles on the whole continent, including North Africa, was estimated at about nine million, 38 percent of which were commercial vehicles (mostly buses, trucks and tractors).[3] A (rather rough) estimate for the early 1980s was that an inhabitant of Africa travelled 100 km per car annually on average, compared to 250 km by bus or lorry and 25 km by train – but 2,000 km on foot (Moriarty and Beed 1989: 126). In 2010, one generation later and after the onset of the African economic renaissance at the turn of the millenium, market observers estimated the total number of motor vehicles in Africa at 26.6 million, of which 9.1 million were commercial vehicles and 17.5 million were cars.[4]

Statistics for African countries indicate that on the average the number of commercial vehicles is significantly over one third of all vehicles. In some countries this number is equal or even higher than the number of private cars. (It should be noted that this is merely a statistical distinction, which is not necessarily valid on the ground. A huge number of private cars is used for commercial purposes, with or without licence.) The comparatively high proportion of commercial vehicles supports our argument that the private car driven

3 Excluding South Africa with its already high rates of motorisation, the number of cars in Sub-Saharan Africa was estimated at 3.6 million and the number of commercial vehicles at 2.2 million (Marchées tropicaux et méditerranéens 1983: 2735).

4 If figures for North Africa are deducted, the number of motor vehicles in Sub-Saharan Africa can be estimated at 15.8 million – as against slightly over 1 billion worldwide (in 2014). Figures are according to Ward's numbers of vehicles in operation by country: http://wardsauto.com/ar/world_vehicle_population_110815 (accessed January 5, 2015). Note that Ward's lists numbers for registered vehicles which in some cases might widely differ from numbers of vehicles actually on the road (see also Beck, this volume).

by the solitary driver might be the ideal realised by an elite, but the reality of ordinary African travel is collective. In rural Africa in particular, the daily car commuter that dominates North Atlantic travel is virtually absent.

The number and density of motor vehicles also vary widely across Sub-Saharan Africa. They tend to mirror the state and the length of the respective road networks of different countries – being highest in southern Africa and lowest in the lower income countries.[5] Like roads themselves, motor vehicles are heavily concentrated in urban areas and on the main road corridors, while in upcountry areas, ratios of one motor vehicle per 10,000 inhabitants are common in some countries. In rural areas 'roadlessness' still goes with 'car famine' – 'une disette d'automobiles', as the editors of Marchées tropicaux et méditerranéens (1983: 2735) put it. As strange as it may seem from the limited perspective of African central business districts, mobility across large areas of Africa is restricted to walking or, at most, to the market lorry drive.

We have made an effort to do justice to these huge variations and associated experiences in the present volume. Bize's account of traffic jams in Nairobi – the only one dealing with urban roads in a volume otherwise dedicated to the African long-distance road – and Stasik's account of the Accra overland bus station both capture the experience of overcharged high-density road environments. Bize in particular highlights the dark side of the road in concentrating on Nairobi's jams (it is ironic that her contribution that deals with well-maintained roads is also at the heart about immobility). On the other hand, Musch's and Wadelnour's contributions on driving a desert route on the trail of African migrants to Europe and on the Sudanese Forty Days Road, respectively, provide accounts of road use amidst a walking or, in these cases rather, a riding world.

The Road of the State

Living in a walking world leads to poor physical access and poor physical access in turn breeds poverty and disease, as runs a well-known line of argument. Off-road communities suffer from a lack of access to health care, to trade and

5 Today, in the rich North Atlantic countries, the number of motor vehicles per 1,000 inhabitants ranges from six to eight hundred motor vehicles. In South Africa, figures rose from 109 per 1,000 inhabitants in 1980 to 165 in 2007; in countries such as Ghana, Nigeria, Senegal and the Sudan, from around 10 to 15 motor vehicles per 1,000 inhabitants in 1980 to between 20 and 30. Lowest vehicle densities are recorded for countries such as Chad (2.6 motor vehicles per 1,000 inhabitants in 1980 and 6 in 2007) and Ethiopia (1.7 motor vehicles per 1,000 inhabitants in 1980 and 3 in 2007).

markets, to credit and banks, as well as from low levels of political and social participation (Porter 1995, 2002; Bryceson et al. 2008). The remedy here is obviously road construction and maintenance. There may, however, be a downside to road building because roads do not bring all-round development, but rather they tend to reshape the social space of a region – a process that usually produces winners as well as losers. The movements of populations to newly constructed roads are universal, with consequences that services located in off-road settlements become increasingly impoverished or even relocate to the road altogether. In turn, this means that the old settlements themselves become deserted and decaying (Porter 1995).

The developmental state derives a good deal of its legitimacy from road construction and the general 'enchantments of infrastructure'. The imaginations of a well-engineered road – with all the promises of modernity and of taming an unruly nature that come with it – have a capacity to fascinate beyond the roads' purely pragmatic effects (Harvey and Knox 2012). On the other hand, there are roads, as depicted in Musch's and Wadelnour's accounts, which have never received road construction from the state, but have been made solely by their users. They fall under the international classification of 'unclassified' or even 'undesignated', the 'orphans' in World Bank parlance, regardless of their importance in terms of value transported on them. The ubiquitous 'votes for roads' strategies in African politics and the road as the beef brought home by local politicians is hinted at in Lamont's contribution, whereas Klaeger's contribution illustrates the workings of the political imagination related to road construction on the Accra-Kumasi road.

It is probably inevitable that the state, with its two-faced nature, not only provides particular services but also acts as a means of political control and coercion. Roads make large tracts of land accessible to those in power, and open them up for processes of 'primitive accumulation'. The same applies to remote populations who come into the range of their state's administration after the building of a road (see Gewald, Luning and van Walraven 2009: 5). Inaccessibility, which may entail remoteness and poverty, can also be interpreted as a shield for local autonomy. From this perspective therefore, poor access might also lead to weak control, and roads may be viewed as 'paths of authority' and not necessarily as paths of development (Fairhead 1992). Consider for instance Fairhead's account on roadbuilding in Zaire's Kivu region, where 'men are in fact afraid to walk along roads […] because they are likely to be robbed there by political cadres, the police of the Mwami and Chef de Groupement, the gendarmes, and the soldiers' (Fairhead 1992: 22). Here robbery is reported to be 'often overt, although it can also be camouflaged through trumped-up charges' (Fairhead 1992: 22). The North Atlantic concept that the modern state guarantees the right of free movement is only applicable to those protected from such extortion by those wielding the state's powers, and who thus view roads in a more positive light – and certainly not in cases of states with a long history of forced exaction like Zaire.

The state is not necessarily the guarantor of the open road in Africa, and the order of the road then seems largely determined by practices beyond the reach of the official state. In fact, roadblocks frequently serve a wide range of policemen, soldiers, militias and/or highwaymen, both legal and illegal, as means to tax traffic passing through their territories (Beek 2011; Lombard 2013). On the whole, Africa has the worst record worldwide in the trucking industry in terms of safety and transit time as well as extortion on the road (Bako-Arifari 2006), and all three should be regarded as interrelated. As in the case of road and vehicle density, variations between the different regions are high. According to a World Bank report (Teravaninthorn and Raballand 2009: 38) transit time from Cotonou to Niamey (1,000 km) has been estimated as six to eight days and from Douala to Ndjamena (1,800 km) as 12 to 15 days. The authors of the report believe that transit times and prices depend less upon road quality and more on what could be called the politics of delay at ports, customs stations and roadblocks (2009: 7). In some regions, roads have developed into regular battlefields for practices of extortion, with some having become quite notorious, such as the highway between Nyala and El Fasher in Darfur, where no less than forty roadblocks were recorded in August 2014 for a total distance of less than 200 km (Sudan Tribune, 5 August 2014). From this and similar experiences we are tempted to propose roadblocks per 1,000 km of road as an enlightening ratio for cross-country comparisons in transportation statistics. In some regions, such common politics of delay probably yields even higher taxes than the official taxes on fuel. Along the central and western African main transport corridors, costs reported for bribes for instance are generally higher than maintenance costs for trucks, rising to more than one quarter of all variable costs on some central African road corridors. Bribes along the eastern and southern African road corridors are reportedly generally lower (Teravaninthorn and Raballand 2009: 69). As a result, law enforcement on the road is low – even on highly supervised highways, as illustrated in Beck's contribution on the Artery of the North highway in the Sudan. Low law enforcement, whether in regard of roadworthiness of vehicles or drivers' adherence to traffic regulations together with bribery, easily translates into the infamous carnage on African roads (Lamont 2010). Some sources believe that Africa has ten percent of the world's road fatalities with less than four percent of the world's vehicle fleet (Gwilliam 2011: 37).

Beyond arguments on the developmental and coercive state, state building and road building are seen as intimately connected on a more fundamental level of order. Whether the purpose is plunder or welfare, roads pave the way for state intervention and they are thus a vital element in the state's territorialising project (Wilson 2004). This connection is usually viewed from the angle of the state, and the argument runs that the state constructs roads for administrative and integrative needs: states make roads. There have indeed been cases of roads in Africa with such a scarcity of traffic that road construction can only be explained as a will to govern and to impress the subjected population with tangible proof of the ruler's paramount power to shape the world, particularly in the early colonial period (Alber 2002; von Trotha 1994: 76). Ciabarri (this volume) explicitly takes a reverse perspective, which is unique in this context as he argues that the road made the state in Somaliland out of the ruins of Somalia's disintegration. He further argues that it is this state-making affordance of the road, which he refers to as 'road effects', that explains the nature of the Somaliland state as a commercial state. This argument about the relation between mobility, power and space could be taken further. Increased mobility grants advantages in any contests about space, whether regarding state or non-state actors, as von Clausewitz has already noted (1991 [1832]). This is clearly spelled out in Klute's (2013: 424–523) seminal work on the Tuareg rebellion (especially in his chapters 'Small wars and space' and 'Small wars and time') and in Musch's piece (this volume) on the Teda's success in taking over strategically important territory from the Tuareg by virtue of their fast desert-going vehicles.

African Automobilities?

An expanding body of scholarship on the road and the automobile has emerged with the so-called mobility turn in the social sciences since the turn of this century.[6] 'Automobility' has been the most influential concept with which the study of traffic and roads has transcended the rather limited realm of planning and engineering and found its way into mainstream social sciences. According to the main exponents of the concept, automobility denotes mobility which is 'auto' in a double sense: self-propelled and autonomous. Central to it is the private car and the freedom of the road (Featherstone 2004; Urry 2004).

The sociologist John Urry conceives of automobility as an autopoietic system. After having been put into motion at the end of the 19th century, this system – like Weber's iron cage – irreversibly reproduces the preconditions

6 For an overview of central (particularly anthropological) works within this body of scholarship, see Dalakoglou (2010), Dalakoglou and Harvey (2012) and Klaeger (2013a).

of its global self-expansion in a path dependent pattern of development. For instance, private car ownership fosters sub-urbanism, commuting, and large retail centres at urban peripheries, which in turn require car ownership to participate in society. 'Social life', Urry writes, 'was irreversibly locked in to the mode of mobility that automobility generates and presupposes' (2004: 27). Accordingly, automobility embraces and connects six components: the car as (i) 'the quintessential manufactured object produced by the leading industrial sectors of the 20th century', (ii) 'the major item of individual consumption after housing which provides status to its owner/user through its sign-values', (iii) a 'powerful complex constituted through interlinkages with other industries', with a connected infrastructure and settlement patterns, (iv) 'the predominant global form of "quasi-private" mobility that subordinates other mobilities', (v) 'the dominant culture that sustains major discourses what constitutes the good life', and at the same time (vi) the 'single most important cause of environmental resource-use' (Urry 2004: 25–26). Automobility, with its promises of the autonomous self, of speed, flow and connectivity has undoubtedly become a *fait total social* of modernity and the freedom of the road has de facto been elevated into the status of a basic right.

The concept of automobility has since then been critically reviewed. It has been criticised for its homogenising tendency at the expense of alternative mobilities such as cycling, sailing, travelling by railway or walking (Vannini 2009). It has also been criticised for its systems theory approach at the expense of political agency and for its own internal contradictions, such as its bleak prospects in terms of ecological sustainability (Böhm et al. 2006). Not least, the concept has been questioned due to its Euro- and America-centric bias, for instance by historians of automobility in the USSR and its satellites (Siegelbaum 2008, 2011).

To conceptualise automobility worldwide as based upon the autonomous driver of the private car would be a gross universalisation of a particular North Atlantic experience, which arises from vehicle densities of 600 to 800 motor vehicles per 1,000 inhabitants (particularly in the United States where private car use tends to be the only form of mobility). Such a perspective certainly does not apply to places with vehicle densities below ten per 1,000 inhabitants, as in a number of African countries that do not share the historical experience of mass (auto)mobility, and in all likelihood never will. The African experience of mobility is that the private car is available only for an elite. The majority of people are transported and probably spend as much time waiting for transport as they do being transported, or they inhabit a 'walking world' altogether. The autonomous driver may be there but the solitary driver is certainly a very rare species.

The relationship between the road and the car in Africa differs from the North Atlantic relationship. The crucial question here is: who pays for the road? In North Atlantic automobile history (with some notable exceptions) cars have paid for the road through fuel tax. African roads, however, are constructed from general government budgets, often by way of public debt, and by international donors. Fuel, instead of being taxed, is even subsidised in a number of countries. Rural and feeder roads are not even paid at all but emerge through use. Thus a different path to road network expansion emerges in the case of Africa in the same way as a different path has been reported in the case of the USSR, where roads have been constructed in large campaigns using corvée labour, POW labour and even the Red Army (Siegelbaum 2011). We therefore suggest that the North Atlantic relationship between the road and the car is rather unique, and certainly not universal. In low-income African countries, as much as half of the expenditure for road construction is donor financed – but not road maintenance (Foster and Briceño-Garmendia 2010: 218), which accounts for much of the state of the road in Africa, as described above. In these cases, it would be an error to regard automobility as a sustainable system, let alone a self-expanding or even self-perpetuating one.

The African road therefore differs significantly from North Atlantic road regimes and corresponding conceptualisations of 'automobility'. In the North Atlantic model, road making is mainly top-down in the style of monopolised Large Technological Systems. Road making is in the hands of planning bureaucracies, administrators and policy makers who, despite being accountable to a public, regulate modes of interaction on and alongside the road through laws, penalties and the built environment. African road regimes, in turn, are to a large extent negotiated by road users in a bottom-up manner in the style of 'inverse infrastructure building' (Egyedi and Mehos 2012); i.e. through practices and interactions of roadside and travel communities. Extensive sections of road infrastructure in the hinterland, for instance Sudan's desert tracks, have come into existence 'under the wheel' without any planning bureaucracy, as Beck's account in this volume illustrates. Others have literally been taken over by self-organised roadside communities in a situation of retreating stateness and deteriorating roads (Lamont, in this volume). The majority of roads have come into existence and have been maintained through the initiatives of local entrepreneurs and traditional authorities, and with the help of local communities. In the case of Ghana, popular initiatives for connecting rural and off-road communities to trunk roads are described in Hill's (1963) account of cocoa farming communities, in Wrangham's (2004) piece on the 'African road revolution', as well as in Klaeger's contribution. In this sense, it is the communities of practice on and along the road that largely regulate and co-create the socio-technological orders of the road.

Despite claims of universal validity for the car regime in global automobility, the history of automobility based on the private car is so far mostly a western experience confined to some capitalist countries. African travellers rather share the experience of having to search for collective and usually crowded transport on buses, in dala-dalas, matatus, SOTRAMAS, in the back of pickups, in overcrowded *taxis-brousse*, on motorcycle taxis and on top of loaded lorries.

Road Skills and Experiences

From the perspective of people living in an environment of mass automobility – which is captured in the automobilities paradigm – road travel has become so habitual that it appears almost natural. African roads show that this is a learning achievement and not natural at all. Beck, for instance, has met people who travelled in a car for the first time in their lives. In particular he recalls a man in his sixties who was a renowned camel racer in his youth and thus certainly well acquainted with rough movement and speed. This man clumsily boarded his first car in late 1999 (probably much like Beck first mounted a camel in 1980) clinging to Beck's shoulder in bewilderment when the car gathered some speed, and after a kilometre or two of slow driving in the desert alighted from the car thoroughly seasick. Having been a caravan man and having worked as a herder for fifty years, his kinaesthetic sense was evidently developed to walking and riding but not to driving. He clearly felt as ill as somebody might when being on a boat in the sea for the first time. This reminds us of what Wolfgang Schivelbusch (1986) has written about the new sensory experiences and skills of early railway travel.

For most Africans riding on buses and lorries today, the sensory experience of road travel may not be novel, but nevertheless quite intense. What passengers frequently comment upon are the rough rides on bad roads and in dilapidated and crammed vehicles. Further causes for complaint and 'suffering' are episodes of prolonged and exhausting standstills (whilst waiting at bus stations, in traffic jams or at roadblocks) or, once on the road, of excessive and dangerous speeding by impatient drivers. The commentaries and associated imageries may appear simplified and overused, yet they hint clearly at passengers' sensory and emotional engagement with the distinct materiality, physical proximity and temporalities that prevail on the road and in vehicles.

Similar forms of experience are made by roadside residents in Africa who are found to live, work, walk, chat and observe in the immediate vicinity of long-distance roads and traffic – a rather unusual occurrence in North Atlantic contexts. As Klaeger (2013b) has shown for roadside dwellers in Ghana, the everyday practices of residents can be seen as ways of 'spatialising' the road in a de Certeauian sense, of 'reappropriating', domesticating and tangibly experiencing road space (see de Certeau 1984: xiv). The daily work of hawkers who sell their wares alongside long-distance traffic, for instance, is characterised by various emotional experiences. They occur among the roadside vendors in the course of their kinetic interactions with passing cars and passengers and with their differential speeds, which may trigger joy and joviality as well as fear and anger. These emotions are usually also connected to their own and their co-vendors' daring and risky bodily movements, which, to the outside observer, are reminiscent of the long-rehearsed precision of a ballet ensemble (Klaeger 2012b).

What is exhibited in the contexts described above are the skills developed by roadside vendors of habitually weaving in and out of flowing traffic with unerring confidence and of dealing with customers in a highly sensory, communicative and often speedy manner (Klaeger 2012a; Bize in this volume). Various other sets of skills can be observed among the road and roadside communities explored in the contributions to this volume. For instance, ticket sellers deploy their own skills when hunting for passengers in the dense and fast-changing environment of a bus station (Stasik). Even the seemingly inactivity of passengers waiting and observing the loading of the Sudanese lorry involves skilful practices for securing the desired place on top of the load (Wadelnour). Whether on desert tracks or on paved highways, such skills are acquired through long-term practice and experience. This becomes particularly obvious in the case of road transport professionals such as the commercial Teda drivers (Musch). It is their profound knowledge of desert topography and orientation – acquired through their long-standing experience as nomadic and mobile people and grafted onto motorised traffic – that the Teda rely upon when working towards a successful trip.

Making the Road

'Road works' are at the core of this volume. What we mean by road works are the manifold ways in which people work, make and thereby co-produce the African road through road-related practices, interactions and skills. As argued earlier, road making in Africa does not only encompass physical road construction and maintenance that are realised by a variety of actors such as the state, planners, engineers, donors as well as local actors and communities. The making of the African road also encompasses the ways in which road traffic is organised and enabled by different bus station workers (Stasik) or the ways in which desert roads are 'made' through the collective efforts of drivers and passengers in the process of travelling (Wadelnour).

This argument of popular involvement in the making of the road can also be extended to the material construction of the road. It is true that modern highways and other paved roads are constructed by road construction companies commissioned by ministries or through public-private partnerships. This certainly pertains to Africa's network of main, secondary and urban roads estimated at 1.2 million km length. But when it comes to the much larger network of tertiary, unclassified and 'undesignated' roads, which is rather characteristic of Africa's rural areas, it should not be taken for granted that these roads have been constructed in the same way. Countless of these more humble roads, including feeder roads and tracks but also important long-distance roads, have come into existence by local initiative of villagers in an effort to be connected to the next town, market or hospital or simply 'under the wheel' (Wadelnour, Lamont, Beck, all this volume). We maintain that the majority of African roads have in this sense been 'constructed' by local initiative and labour, even if these roads have subsequently been upgraded by construction machinery. And it is mainly because of our North Atlantic experience of road building that we tend to begin an account of road building with the decision in a ministry and the arrival of heavy machinery. In these accounts, the pioneers of the road are all too easily forgotten. A telling parallel from South America is related by Penny Harvey and Hannah Knox about the local pioneers of the Iquitos-Nauta highway in Peru who first opened the way through the dense forest with their machetes in the 1960s and were, of course, sidelined and forgotten when the president ceremonially inaugurated the paved highway in 2006 (Harvey and Knox 2015: 42, 188).

In other contexts, people make the road through their productive use of road space, such as when roadside vendors take advantage of traffic jams and slowed traffic and thus turn the road into a market space (Bize). In yet another dimension, the road is produced by means of specific narratives and commentaries (Klaeger) and as part of people's imaginary and cosmological work. Bürge, for instance, shows that different levels of road making become interlocked when the movements of commercial motorbike riders in town and their presence at crossroads are constitutive of the ways in which the community makes sense of the cosmological substructure of the crossroads.

What we argue, then, is that the making and fabricating of the African road is to be understood in more than one sense. Beyond material construction, road making denotes practices that pertain to, among others, commercial, organisational, spatial and often bodily skills. Yet road making also comprises the communicative construction of the road through narrative, imaginary and cosmological means. Road making therefore embraces a variety of actors who – through their respective efforts and routines – realise the ordering, regulation and signification of the road. What we aim at demonstrating in particular is that the African road emerges as a co-production of different communities of practice and that, in many contexts, it is constituted by the specific interactions of travel and roadside communities.

This perspective illustrates that the road in Africa is not merely to be seen as a problematic case, one which is marked by deficiency when compared to the road in North Atlantic contexts. First and foremost it is to be seen as a human achievement and, in view of its particular characteristics, as a unique phenomenon that merits being paid full attention to by social scientists. As we have discussed earlier (and as the contributions to this volume show), a central characteristic is its differentiated and multifunctional space that it provides to everyday users. This clearly distinguishes it from North Atlantic roads, as it opens up spaces that allow for far more negotiation of its uses and significations than the North Atlantic version does. Further yet, much of what is commonly understood as 'automobility' is built on particular North Atlantic experiences, especially on the assumption of the solitary drivers. If there is something like African automobility, then some of its defining characteristics would be the collective nature of travel; the largely artisanal and informal organisation of transport; as well as the ample opportunities for interactions between travel and roadside communities.

Thus, if we have been able to present a more differentiated perspective on roads and mobility, one which broadens and overcomes the limits of the prevalent North Atlantic perspective on automobility through our portrayal of the African road, then this volume has served its purpose.

Itinerary through This Volume

The contributions to this volume meet at the intersection of anthropological interests with both historical and political science interests. They range from a micro-experiential perspective to a technological perspective and from a narrative or cosmological perspective to a political perspective, some combining several of these perspectives. In the end, they all work towards the idea of the making of the African road.

We have organised the nine contributions as a journey along different types of African roads and roadsides. Our journey leads us from the paved and often jammed roads in urban centres to well-developed inter-regional roads running through densely settled areas and through to hinterland roads in less populated areas which eventually split into unsurfaced tracks and lanes.

In the chapter following this introduction, then, Michael Stasik explores a central bus station in Ghana's capital Accra, where travel communities are formed and 'channelled' onto the roads by various roadside communities. Expanding on the notion of 'involution', Stasik discusses the practices and institutional arrangements by which the communities of the station accommodate the complexities of one of West Africa's busiest road travel hubs.

In Chapter 3, Amiel Bize takes us to the roads in downtown Nairobi, Kenya, zeroing in on the phenomenon of traffic jams. She assesses the role of jammed urban roads from two perspectives: firstly as sites of obstructed flows, halted productivity and exhaustion that induce governance rhetorics about regulation and control, and secondly as actual worksites of informal roadside vendors whose practices are honed in on making the seemingly 'wasteful' spaces and temporalities of clogged roads productive.

Chapter 4 by Gabriel Klaeger focuses on road-related social commentaries and imaginations of people living in Kyebi, Ghana, a town that has been detached from a main interregional road through the recent construction of a bypass. By examining the conflicting stories and rumours triggered by the 'removal' of the road, Klaeger shows how the road is perceived and narratively constructed as an ambivalent realm of 'power, progress and perils'.

Common wisdom has it that the state promotes administrative and economic capacities through the development of road infrastructure. Luca Ciabarri in Chapter 5 provides a contrasting perspective. Framed by an analysis of the political economy in the Horn of Africa, he delineates how a commercial corridor, comprising a port, a road and free trans-border movement, was instrumental in the development and 'making' of the state of Somaliland.

Chapter 6 by Michael Bürge throws into relief the cosmological substructure of roads. Centred on a crossroads in the town of Makeni, Sierra Leone, he describes the ambivalent role of commercial motorbike riders, the *Okadamen*. Representing both agents of mobility and connection and mongers of potentially malign spirits, Bürge suggests, the *Okadamen* embody the liminal, 'transmogrifying' powers of the crossroads they inhabit and work.

After the Sierra Leonean crossroads, we leave behind paved roads and continue on the dirt tracks of rural areas. Thus, Chapter 7 by Mark Lamont brings us to the unimproved road infrastructure in the Kenyan hinterlands. By way of an ethnography of some fifty kilometres stretch of road, Lamont describes how instances of infrastructural tension, failure and 'ruin' bring into being particular forms of sociality centred on maintenance and repair.

In Chapter 8, Rami Wadelnour introduces us to the desert tracks of the Forty Days Road in western Sudan. In his account, it is the traffic that brings the road into existence. Wadelnour describes these popular ways of road making by attending to the collective practices of both drivers and passengers who use and thus 'produce' the road.

Chapter 9 by Tilman Musch takes us to the border region of Niger, Chad and Libya. He describes how a former Tuareg desert caravan route has been taken over by commercial Teda drivers who transport migrants on their way to Europe. The Teda's success, Musch argues, is facilitated primarily by their combination of long-established knowledge of desert topography and motorised means of travel and road making.

Having reached the vast and seemingly 'roadless' spaces of the desert, we do not leave the tracks to peter out into no man's land but close with a perspective that embraces the different orders of the roads we have travelled on. In Chapter 10, then, Kurt Beck exemplarily contrasts with each other two distinct yet interlinked technological regimes of road travel in northern Sudan. Expanding on the notion of 'technological dramas', he examines how the orders of a modern highway regime become established through contesting claims by state authorities and users of an older road regime of desert travel.

References

Alber, Erdmute. 2002. 'Motorization and Colonial Rule: Two Scandals in Dahomey, 1916.' *Journal of African Cultural Studies* 15 (1): 79–92.

Augé, Marc. 1992. *Non-Lieux: Introduction a une anthropologie de la surmodernité*. Paris: Éditions du Seuil.

Bako-Arifari, Nassirou. 2006. 'We Don't Eat the Paper: Corruption in Transport, Customs and the Civil Forces'. In *Everyday Corruption and the State: Citizens and Public Officials in Africa*, edited by Giorgio Blundo and Jean-Pierre Olivier de Sardan, 177–224. London: Zed Books.

Beck, Kurt. 2013. 'Roadside Comforts: Truck Stops on the Forty Days Road in Western Sudan.' *Africa* 83 (3): 426–445.

Beek, Jan. 2011. '"Every Car has an Offence on it": Register polizeilichen Handelns bei Verkehrskontrollen in Nordghana.' *Sociologus* 61 (2): 197–222.

Bijker, Wiebe, Thomas Hughes and Trevor Pinch (eds). 1989. *The Social Construction of Technological Systems: New Directions in the Sociology and History of Technology*. Cambridge: MIT Press.

Böhm, Steffen, Campbell Jones, Chris Land and Matthew Paterson (eds). 2006. *Against Automobility*. Cambridge: Blackwell.

Bryceson, Deborah, Annabel Bradbury and Trevor Bradbury. 2008. 'Roads to Poverty Reduction? Exploring Rural Roads' Impact on Mobility in Africa and Asia.' *Development Policy Review* 26 (4): 459–482.

Dalakoglou, Dimitris. 2010. 'The Road: An Ethnography of the Albanian-Greek Cross-Border Motorway.' *American Ethnologist* 73 (1): 132–149.

Dalakoglou, Dimitris and Penny Harvey. 2012. 'Roads and Anthropology: Ethnographic Perspectives on Space, Time and (Im)Mobility.' *Mobilities* 7 (4): 459–465.

Dant, Tim. 2004. 'The Driver-Car.' *Theory, Culture and Society* 25 (4/5): 61–79.

de Certeau, Michel. 1984. *The Practice of Everyday Life*. Berkeley: University of California Press.

Edensor, Tim. 2004. 'Automobility and National Identity: Representation, Geography and Driving Practice.' *Theory Culture Society* 21 (4/5): 101–120.

Egyedi, Tineke and Donna Mehos (eds). 2012. *Inverse Infrastructures: Disrupting Networks from Below*. Cheltenham: Elgar.

Fairhead, James. 1992. 'Paths of Authority: Roads, the State and Market in Eastern Zaire.' *The European Journal of Development Research* 4 (2): 17–35.

Featherstone, Mike. 2004. 'Automobilities: An Introduction.' *Theory, Culture and Society* 21 (4/5): 1–24.

Foster, Vivien and Cecilia Briceño-Garmendia. 2010. *Africa's Infrastructure: A Time for Transformation*. Washington: IBRD.

Gewald, Jan-Bart, Sabine Luning and Klaas van Walraven. 2009. 'Motor Vehicles and People in Africa: An Introduction.' In *The Speed of Change: Motor Vehicles and People in Africa, 1890–2000*, edited by Jan-Bart Gewald, Sabine Luning and Klaas van Walraven, 1–18. Leiden: Brill.

Godard, Xavier. 2002. *Les transports et la ville en Afrique au sud du Sahara: Le temps de la débrouille et du désordre inventif. Économie et développement*. Paris: Karthala.

Godard, Xavier and Pierre Teurnier. 1992. *Les transports urbains en Afrique à l'heure de l'ajustement: redéfinir le service public. Villes et citadins*. Paris: Karthala.

Gwilliam, Ken. 2011. *Africa's Transport Infrastructure: Mainstreaming Maintenance and Management*. Washington: World Bank.

Gwilliam, Ken, Vivien Foster, Rodrigo Archondo-Callao, Cecilia Briceño-Garmendia, Alberto Nogales and Kavita Sethi. 2009. *The Burden of Maintenance: Roads in Sub-Saharan Africa*. Washington: World Bank.

Harvey, Penny and Hannah Knox. 2012. 'The Enchantments of Infrastructure.' *Mobilities* 7 (4): 521–536.

Harvey, Penny and Hannah Knox. 2015. *Roads: An Anthropology of Infrastructure and Expertise*. Ithaca: Cornell University Press.

Hill, Polly. 1963. *The Migrant Cocoa-Farmers of Southern Ghana: A Study in Rural Capitalism*. Cambridge: Cambridge University Press.

Klaeger, Gabriel. 2009. 'Religion on the Road: The Spiritual Experience of Road Travel in Ghana.' In *The Speed of Change: Motor Vehicles and People in Africa, 1890–2000*, edited by Jan-Bart Gewald, Sabine Luning, and Klaas van Walraven, 212–231. Leiden; Boston: Brill.

Klaeger, Gabriel. 2012a. 'Rush and Relax: The Rhythms and Speeds of Touting Perishable Products on a Ghanaian Roadside.' *Mobilities* 7 (4): 537–554.

Klaeger, Gabriel. 2012b. 'Movements into Emotions: Kinetic Tactics, commotion and Conviviality among Traffic Vendors in Accra.' In *Urban Life-Worlds in Motion: African Perspectives*, edited by Hans Peter Hahn and Kristin Kastner, 131–156. Bielefeld: transcript.

Klaeger, Gabriel. 2013a. 'Introduction: The Perils and Possibilities of African roads.' *Africa* 83 (3): 359–366.

Klaeger, Gabriel. 2013b. 'Dwelling on the Road: Routines, Rituals and Roadblocks in Southern Ghana.' *Africa* 83 (3): 446–469.

Klute, Georg. 2013. *Tuareg-Aufstand in der Wüste: Ein Beitrag zur Anthropologie der Gewalt und des Krieges*. Köln: Köppe.

Lamont, Mark. 2010. 'An Epidemic on Wheels? Road Safety, Public Health and Injury Politics in Africa.' *Anthropology Today* 26 (5): 3–7.

Larkin, Brian. 2013. 'The Politics and Poetics of Infrastructure.' *Annual Review of Anthropology* 42: 327–343.

Lombard, Louisa. 2013. 'Navigational Tools for Central African Roadblocks.' *Political and Legal Anthropology Review* 36 (1): 157–173.

Marchées tropicaux et méditerranéens. 1983. *L'automobile en Afrique* (Special Issue). *Marchées tropicaux et méditerranéens* 39 (1984): 2715–2790.

Marx, Karl. 1968. *Das Kapital: Kritik der politischen Ökonomie*. MEW Vol. 23. Berlin: Dietz Verlag.

Mauch, Christof and Thomas Zeller (eds). 2008. *The World Beyond the Windshield: Roads and Landscapes in the United States and Europe*. Athens: Ohio University Press.

McCarthy, Tom. 2007. *Auto Mania: Cars, Consumers, and the Environment*. New Haven & London: Yale University Press.

Merriman, Peter. 2004. 'Driving Places: Marc Augé, Non-Places, and the Geographies of England's M1 Motorway.' *Theory, Culture & Society* 21 (4/5): 145–167.

Merriman, Peter. 2007. *Driving Spaces: A Cultural-Historical Geography of England's M1 Motorway*. Malden: Blackwell.

Moriarty, Patrick and Clive Beed. 1989. 'Transport in Tropical Africa.' *The Journal of Modern African Studies* 27 (1): 125–132.

Porter, Gina. 1995. 'The Impact of Road Construction on Women's Trade in Rural Nigeria.' *Journal of Transport Geography* 3 (1): 3–14.

Porter, Gina. 2002. 'Living in a Walking World: Rural Mobility and Social Equity Issues in Sub-Saharan Africa.' *World Development* 30 (2): 285–300.

Schivelbusch, Wolfgang. 1986. *The Railway Journey: The Industrialisation of Time and Space in the 19th Century*. Berkeley: University of California Press. (German edition: 1977. *Geschichte der Eisenbahnreise*. München: Carl Hanser Verlag.).

Sheller, Mimi. 2004. 'Automotive Emotions: Feeling the Car.' *Theory, Culture and Scociety* 21 (4/5): 221–242.

Siegelbaum, Lewis. 2008. *Cars for Comrades: The Life of the Soviet Automobile*. Ithaca: Cornell University Press.

Siegelbaum, Lewis. (ed.). 2011. *The Socialist Car: Automobility in the Eastern Bloc*. Ithaca: Cornell University Press.

Star, Susan. 1999. 'The Ethnography of Infrastructure.' *American Behavioral Scientist* 43 (3): 337–391.

Stasik, Michael. 2013. 'In the Hustle Park: The Social Organisation of Disorder in a West African Travel Hub.' *Working Papers of the SPP 1448*, Halle and Leipzig, Nr 1.

Stoller, Paul. 1989. *The Taste of Ethnographic Things*. Philadelphia: University of Pennsylavania Press.

Teravaninthorn, Supee and Gaël Raballand. 2009. *Transport Prices and Costs in Africa: A Review of the Main Corridors*. Washington: World Bank.

Thrift, Nigel. 2004. 'Driving in the City.' *Theory, Culture and Society* 21 (4/5): 41–60.

Urry, John. 2004. 'The "System" of Automobility.' *Theory, Culture and Society* 21 (4/5): 25–39.

Vannini, Phillip. (ed.). 2009. *The Cultures of Alternative Mobilities: Routes less Travelled*. Farnham: Ashgate.

Volti, Rudi. 1996. 'A Century of Automobility.' *Technology and Culture* 37 (4): 663–685.

von Clausewitz, Carl. 1991 (1832). *Vom Kriege*. Edited by Werner Hahlweg. Bonn: Dümmlers.

von Trotha, Trutz. 1994. *Koloniale Herrschaft: Zur soziologischen Theorie der Staatsentstehung am Beispiel des Schutzgebietes Togo*. Tübingen: Mohr.

Wilson, Fiona. 2004. 'Towards a Political Economy of Roads: Experiences from Peru.' *Development and Change* 35 (3): 525–546.

Wrangham, Elizabeth. 2004. 'An African Road Revolution: The Gold Coast in the Period of the Great War.' *The Journal of Imperial and Commonwealth History* 32 (1): 1–18.

CHAPTER 2

Roadside Involution, Or How Many People Do You Need to Run a Lorry Park?

Michael Stasik

In Ghana, public road transport is not a public undertaking but is largely in the hands of small-scale entrepreneurs, and it has been so since the advent of motorised transportation in the early twentieth century. The related technologies (above all, roads and vehicles) and the very model of motorised mass transportation were imported from the North Atlantic regions, mainly from Britain. During the first decades of what Polly Hill has famously coined as the 'lorry age' (1963b: 234) – which alludes to the great transformations brought about by the motorisation of Ghanaian society since the 1910s –, the 'original' model of public transport got adopted to the cultural syntax of West African practices of travel. In the course of this process, Ghanaian modes of road transport began to 'deviate' significantly from the forms of public transport in the British 'motherland' and the North Atlantic regions at large. In the North Atlantic regions, public transport follows bureaucratically administered models with high levels of regulation, standardisation and formalisation. The European railway systems, for example, epitomise the cogwheel rigidity of Western states' public transport planning and organisation through their centrally-controlled routes, schedules, fares, as well as meticulously differentiated set of scripts devised for both its operators and its users.

Ghana's privately run public road transport, by contrast, is characterised by conspicuously low degrees of central planning and regulation. Being an artisanal 'trade' with decentralised, diverse and bottom-up deployed structures of operations, it is often considered to be emblematic of what is commonly described as the 'informal' economy (e.g. Barrett 2003; Hansen and Vaa 2004; Hart 1970, 1973). The practices of road operators and users do not follow any top-down prescribed set of scripts, rather, its modes of organisation have evolved from distinct repertoires of skilled, often tacit practices and quotidian interactions that take place on the road and roadside. The distribution of routes is not centrally allocated, but according to the dynamics of supply and demand. These routes are served by a broad range of vehicle types; vehicles do not run according to time schedules, but only depart once the last seat has been taken; although fares are fixed according to officially set rates, in practice they are routinely adjusted at will.

 | DOI 10.1163/9789004339040_003

As in most countries across Africa, such forms of informal collective road transport account for the main means of motorised transportation in Ghana (MRH and SSGH 2009; Trans-Africa Consortium 2010). Most inner-city, inter-regional and international travels, in turn, are organised in urban bus stations, called 'lorry parks' in Ghanaian English. This is where travel communities are formed and 'channelled' into their journeys by various roadside communities of the station. Lorry parks are hubs in the full sense of the word: they are pivots of travel and trade, gateways between urban and rural areas and interregional and international intersections. They are kept running through diverse networks of loosely-structured group economic activities, the constituents of which are far from clear to all of its participants. As most road-bound travels begin in urban bus stations, the practices of the station communities – who 'dispatch' passengers and goods, thus bringing traffic onto the roads – figure as central elements in the overall 'making' of road transportation.

With regards to the decisive role of bus stations within the complex operations of public transport in Ghana, and similarly in other African contexts, it is surprising that they have not received much attention from scholars dealing with related phenomena (roads, travel and transportation). This apparent lack of interest is particularly striking in relation to the work of anthropologists. Since the advent of automobility in Africa, many if not most anthropologists working in Africa must have spent considerable amounts of time at bus stations. Sjaak van der Geest, for example, reflecting upon the three and a half decades of his returning visits to the continent states that in 'those years I spent countless hours in lorry parks waiting for my bus or taxi to leave', during which he 'killed time by talking to drivers, "bookmen", mates and fellow passengers' (2009: 269). Van der Geest's long-lasting, albeit not deliberate, bond with African bus stations is certainly not an exception among fellow Africanist fieldworkers both within and outside the scopes of anthropological research.[1] And still, as Paul Nugent has recently noted with reference to the comparably vital locales of African markets, while 'markets have received their fair share of academic treatment, lorry parks have not received nearly enough attention as interactive spaces' (2010: 96).

Most noteworthy among the few studies dealing with Africa's bus stations are Polly Hill's portrayals of Ghanaian trade and transport systems in the 1960s

1 Van der Geest hints at a possible explanation for that 'lorry park-desideratum' by pointing to the 'dismal farsightedness' of (Western) anthropologists: 'What originates in their own culture is too familiar to be visible in another setting: schools, factories, hospitals, pharmaceuticals and cars – all these exports have – until recently – been overlooked by Western anthropologists doing fieldwork in other cultures' (2009: 259).

(1962, 1963a, 1965, 1984) and Paul Stoller's 'deep reading' of social interactions in a bush taxi station in Niger (1989: 69–83). Many other writings related to bus stations appear to have emerged rather as subsidiary products of research on other subjects, with studies on urban transport infrastructures accounting for the largest share.[2] Descriptions of stations can be found scattered in literature dealing with the politics of transport and the transport economy,[3] with trade and market places[4] and with urban lifeworlds more generally.[5] In a number of anthropological (and anthropologically-inclined) studies, the significance of stations is touched upon via discussions of some of its central figures and technologies; mainly commercial drivers and vehicles,[6] as well as touts, hawkers, passengers and itinerant preachers.[7] Finally, a surprisingly large amount of literature deals with car inscriptions, or 'minivan poetics' (Weiss 2009: 49), the collections of which were largely compiled during researchers' 'idle time' at stations.[8]

In this chapter, I expand upon this scattered body of works on African bus stations. In my approach, I put the 'interactive space' of the station at the centre of attention, rather than attending to it parenthetically. I do so by way of a minute exploration of the practices and the 'practitioners' in a long-distance bus station in Ghana's capital Accra, the so-called Neoplan Station. I draw on twelve months of fieldwork conducted between 2011 and 2013, during which I engaged with the roadside communities of the station mainly by means of what Gerd Spittler (2001) coined 'thick participation'. Thus, in addition to the

---

2 There is an enormous amount of policy papers, analyses and reports on African urban transport provisions (and 'problems'). Yet, as these only tend to mention stations, rather than examining their workings, I include references in which inner-city bus stations are dealt with more explicitly and beyond the mere scopes of stakeholder analyses. For Ghana, see Fouracre et al. 1994; IBIS 2005; Kwakye and Fouracre 1998; Turner 1996. For urban stations in other African countries, see Barrett 2003; Coulibaly 1993; Hawkins 1958; Jordan 1983; Khayesi 2002; Lopes 2009; Ndiaye and Tremblay 2009.

3 Albert 2007; Bähre 2014; Cissokho 2012; Czeglédy 2004; Horta 2013; Jalloh 1998; Jennische 2012; Joshi and Ayee 2002; Khosa 1992; Lewis 1970; Peace 1988; van Walraven 2009.

4 Clark 1994; Grieco, Apt and Turner 1996; Klaeger 2012; Ntewusu 2012; Stasik and Thiel 2014.

5 Ademowo 2010; Daniel 2012; Lamote 2007; Simone 2004: 410; Titus et al. 2010.

6 Beisel and Schneider 2012; Clayborne 2012; Feld 2012: 159–197; Hart 2011; Jordan 1978; Klaeger 2014; Mutongi 2006; Olubomehin 2012; Sesay 1966; van der Geest 2009; Verrips and Meyer 2000.

7 Asiedu and Agyei-Mensah 2008; Davies 2008; Klaeger 2009; Mashiri 2001; Okpara 1988; Seck 2006; Tambulasi and Kayuni 2008; wa Mungai and Samper 2006.

8 Collections and interpretations of car inscriptions in Ghana make up the largest part of these works: Awedoba 1981; Date-Bah 1980; Field 1960; Kyei and Schreckenbach 1975; Lewis 1998; Quayson 2010a, 2010b, 2014: 129–158; van der Geest 2009. For Liberia, Nigeria and Tanzania respectively, see Guseh 2008; Lawuyi 1988; Pritchett 1979; Weiss 2009: 45–52.

ethnographic toolkit of observations, interviews and conversations, my research involved sustained participatory experiences that were gathered via apprenticeship and employment in various divisions of the Neoplan Station.

My main aim for this chapter is to complement this volume's focus on the 'production' of African roads from the perspective of a roadside institution that serves as a vital 'feeder' of roads, which is the bus station. I frame this perspective by drawing on the notion of 'involution' (Geertz 1963): an inward-bound process of organisational change (best contrasted with the unfolding changes of 'evolution') that is characterised by increased institutional complexity, 'elaboration and ornateness' and 'unending virtuosity' (Geertz 1963: 82). This kind of 'static expansion' (Geertz 1963: 79, borrowing from Boeke 1953) provides a useful concept for making sense of both the intricate arrangements that structure the workings of the roadside institution of the station as well as the practices that make these arrangements work and that keep the traffic moving up and down the roads.

I suggest that the station's involution and the orders of the roads are constitutive of each other. On the one hand, the station 'feeds off' the road: the more diverse and dense the station-bound traffic is, the more intense and elaborate the trajectories of organisational involution of the station's roadside communities become. On the other hand, the station also 'feeds' the roads. Representing a nerve centre for transportation where travel routes begin, intersect and end, it is in the station that the density and diversity of flows on the roads are enabled, sustained and, at times, also disrupted. It is in here that a large part of motorised road traffic is deployed, maintained and dispatched; that a multitude of road users other than drivers and passengers (hawkers and itinerants, for example) is integrated into a system of transportation; and that the specific rhythmicities evolving on the roads are structured and timed (see Stasik 2015). Ultimately, the involuting structures of the station serve to accommodate, channel and expand the very diversity of road traffic that its own inward-bound expansion co-creates. The perspective from the station highlights that – in Ghana and similarly in other African contexts – roads and roadsides are inextricably linked together.

Accra's Neoplan Station

The Neoplan Station[9] is located near Accra's main road intersection, the Kwame Nkrumah Circle (or simply 'Circle' as it is called in the urban vernacular).

9 The station owes its somewhat odd name to a joint venture that the Stuttgart-based bus manufacturer *Neoplan* initiated with the Ghanaian government in the mid-1970s.

Resembling Ojuelegba in Lagos, Circle functions as 'a vortex for all the flows' (Bakare-Yusuf and Weate 2005: 326) within Accra – a property it owes to hosting Accra's main inner-city bus station (the 'Circle Station'). The Neoplan Station complements Circle's role as a vital transportation hub that supports the overall economies of Accra and Ghana, stemming primarily from Neoplan's role in plying the most important route in the country, that which connects Accra with Ghana's second biggest city, Kumasi.

34 destinations are directly served from Neoplan's yard by 2013. These destinations are mainly scattered across the country's central and western regions and link Accra to the port cities of Tema and Takoradi as well as with Lagos in Nigeria, among others. Neoplan thus acts as Accra's central gateway to all the major commercial centres in Ghana and, through the feeder routes branching off from its destinations, to the West African sub-region at large.[10] Accordingly, the station is frequented by many travellers from diverse backgrounds and with a variety of travel purposes; this includes people from all the regions in Ghana and neighbouring countries as well as from much further afar, such as Senegalese salesmen (and women) traversing overland to Central Africa, southern African migrants on their way to Europe's southern shores or European and American backpackers trekking across the continent.

The numbers of vehicles – and of seated passengers – that are dispatched daily from the station further highlight Neoplan's role as one of Accra's main hubs of travel and transport.[11] Although its daily turnover is subject to considerable fluctuations of weekly, monthly and seasonally conditioned rhythms and to aperiodic perturbations (e.g. heavy rains, major sport events), three main instances can be distinguished: during particularly quiet days (usually

German-produced coachworks and chassis were exported to Ghana and pieced together there, fabricating the *Neoplan Tropicliner*. Well-adapted to the adverse road conditions of the West African tropics, the coach represented the most prestigious type of bus to be found on Ghana's roads. At the opening of the station in 1979, it became its flagship vehicle and name giver.

10 The Neoplan Station is one of three main lorry parks in Accra. These three parks have roughly 'partitioned' major routes in Ghana along three so-called 'corridors': the 'western corridor' served by the Kaneshie Station in Accra's west; the 'eastern corridor' served by the Tudu Station located in downtown Accra; and the 'central corridor', which is served by vehicles from Neoplan. This partitioning of routes is not particularly strict and there are overlaps on significant sections of the corridors between these three stations. Furthermore, many of the routes are also served by vehicles from other, smaller stations in Accra.

11 There is no centralised accounting system at the Neoplan Station. I collated the figures by counting, re-counting and cross-checking the number of cars (with the number of seats and thus passengers) departing from the station over extended periods of time.

Sundays and mid-week), about 10,000 passengers flock into the station's yard; the number doubles on other days of the week to about 20,000 people; and during peak times, as on the last weekend of the month (after payday) and holidays, the number of travellers rises to some 30,000 and more.

Neoplan's fleets of vehicles range from small five-seater saloon cars to old 39-seater Benz buses through to brand-new coaches carrying 50 people. The figures vary between 700 cars on quiet days up to 2,500 departing from the station on exceptionally busy days. Not included in this count are the taxis, delivery trucks and (hand) truck-pushers entering Neoplan's yard every other minute and supplying its 'channelling machinery' with new passengers and goods. Neoplan's yard covers an area of approximately just one hectare, stretching over some 150 by 60 meters. Framed by two L-shaped single-storey constructions that demarcate its spatial boundaries, the yard is the same size as a single railway platform of a secondary city train station in Europe. Measured against its quantity of incoming and departing passengers and vehicles, it is remarkably small and narrow.

Besides being a 'channelling machinery' for the bulk of travellers that are dispatched by hundreds of vehicles, the station attracts all sorts of traders, vendors and hawkers as well as itinerants, sex workers, beggars, thieves and dealers. Its narrow yard accommodates diverse services and institutions, some of which are directly linked to the needs of travel and transport (e.g. workshops and spare parts dealers; shops selling provisions and merchandise; eating parlours and drinking spots), while others have emerged through less direct links (e.g. herbalists, pharmacists and hairdressers; football, cinema and video game centres; a church, mosque as well as the odd insurance company).

The 34 destinations served from Neoplan are run by 13 sub-stations that are locally referred to as 'branches'. These 13 branches form part of larger bodies of nationwide operating transport associations. Neoplan is split between two of the largest associations: ten of its 13 branches belong to the Ghana Private Road Transport Union (GPRTU), whilst the other three branches are affiliated with the Progressive Transport Owners' Association (PROTOA). These associations – which have little in common with the trade unions of employees but rather resemble kinds of guilds – evolved from indigenous drivers organisations in the aftermath of Ghana's (then Gold Coast) 'road revolution' in the late 1920s (Wrangham 2004; see also Hart 2014).[12] Emerging from this long-standing

12 The emergence of the first associations occurred at the same time the first lorry parks were established in the Gold Coast, which the British colonial administration enforced in order to increase the regulative leverage on the motor transport sector. The first main parks opened in Accra, Sekondi and Koforidua in 1929, in 1930 in Kumasi and in 1931 in

FIGURE 2.1–2.4 *A cross-section of Neoplan's vehicle fleet (clockwise from top-left): 'resilient' Lagos Volvo; 'common' Toyota minivan to Nkawkaw; 'well-preserved' Benz bus to Ashaiman (long chassis); '1st class' Kia coach to Kumasi*
SOURCE: M. STASIK, 2012

FIGURE 2.5 *Neoplan's station hawkers, here peddling meat pies, fried rice, biscuits and bread to passengers seated in a departing car*
SOURCE: M. STASIK, 2011

position is their power to control access to the country's main lorry parks, which in effect carries out the only entry controls to the public transport market.

Entry controls are organised on the basis of a vehicle registration system, which is subject to the branches of the park. Vehicles are registered through the payments of a range of fees levied by the branch, parts of which are passed on to regional and national representatives of their respective associations. In principle, there are no prescribed conditions as to what kinds of vehicles qualify (or not) for registering with a branch. In practice, however, it is the kinds of routes that are plied by a given branch – and the respective quality of the roads – that make for a pre-selection. On Neoplan's highly frequented shorter routes, such as the well-developed connection to the neighbouring town of Ashaiman (35 km on a two-lane highway) upon which several hundred commuters are dispatched daily, preference is given to vehicles with a large carrying capacity, with little consideration for the vehicle's condition. Small and 'resilient' cars are preferred on less frequented longer routes, such as the connection to Lagos (500 km, including six border controls) that poses an unpredictable variety of roadway 'irregularities' (e.g. roadblocks, petrol shortages). Some branch members own vehicles themselves, either as 'owner-drivers' or as senior members who commission junior members as drivers. Yet the majority of cars belong to private entrepreneurs who are unrelated to the branches and who hire their cars out to drivers on a commission basis, with usually one vehicle per owner only.[13]

The station branches and their personnel hence serve as brokers between passengers, vehicle owners and the larger bodies of national transport associations. As such, they represent a crucial element in a much-fragmented system of operations, which is as powerful in generating revenues as it is divided over the distribution of gains. These divisions stem primarily from competition over the most lucrative routes that the branches are engaged in. As there is no overall route licencing system, the distribution of routes is structured mainly through market demand and the capacities of each branch to serve those demands and to withstand competitive pressures. The effects of market-driven processes of route organisation are pointedly reflected in the net of itineraries

Cape Coast. The first associations included the *Gold Coast Motor Union* (established in 1929 in Accra), the *Drivers' Association of the Western Province* (Sekondi, 1929) and the *Ashanti Motor Union* (Kumasi, 1930). This suggests a causal rather than a coincidental relationship between the formation of lorry parks and drivers' associations. Archival sources are, however, inconclusive in this regard.

13 The owners of vehicles registered in the Neoplan Station, for example, include civil servants, military personnel, clerks, traders, teachers, college students and a South Korean businessman, among others.

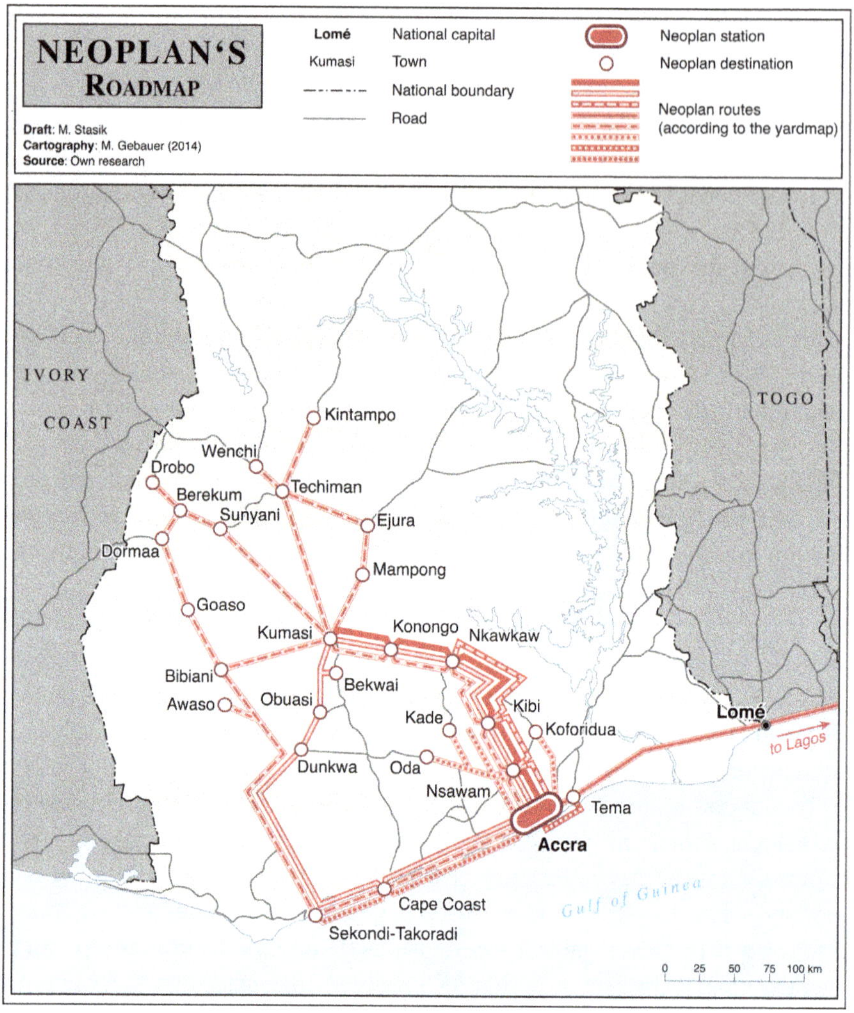

MAP 2.1 *Neoplan's roadmap*
ORIGINALLY PUBLISHED IN STASIK (2015). REPRODUCED WITH PERMISSION FROM TAYLOR & FRANCIS LTD. (WWW.TANDFONLINE.COM).

of the Neoplan Station. Out of its 34 destinations, 19 are served by only one branch (these are the least lucrative routes). The passengers for the remaining 15 destinations are competed for by at least two different branches. The Kumasi-bound cars are in the toughest scramble, as this most travelled route is served by four branches simultaneously. The complexities resulting from these competitive modes of route organisation are both the source and the product of the increasingly complex – and involuting – organisational structures of the branches.

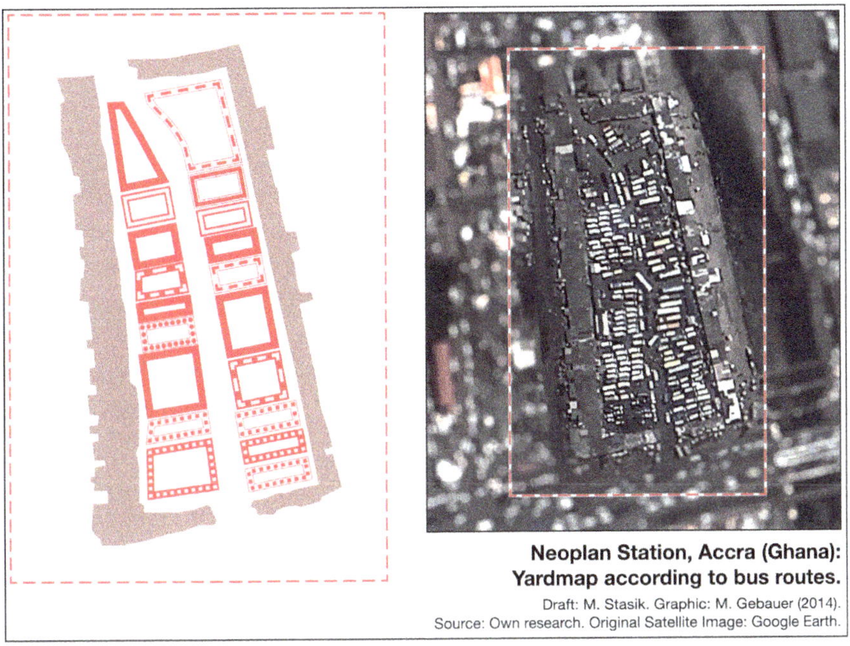

MAP 2.2 *Neoplan's yardmap*
ORIGINALLY PUBLISHED IN STASIK (2015). REPRODUCED WITH PERMISSION FROM TAYLOR & FRANCIS LTD. (WWW.TANDFONLINE.COM).

Yard Involution

Neoplan's 13 branches are made up of 'office staff' and 'yard staff'. The office staff represents each branch's administrative body. It has a board of five 'officers': a chairman, vice-chairman, secretary and two trustees who are elected every four years from members of the branch. The chairman is the main representative of the branch. Addressed as 'the boss' (*panyin*, in Twi) he is the ultimate authority for imposing penalties, disciplining or dismissing branch members, and he acts as the main connection to the regional and national levels of the respective association. Whilst some chairmen capitalise on their positions for obtaining (moderate) wealth and even political influence, most serve as mere ombudsmen, whose re-election is but a matter of routine. Apart from serving as the first deputy, the vice-chairman is mainly in charge of issuing the directives of the chair to the branch members. The main task of the two trustees is the administration of finances. Correspondence, bookkeeping and the formal registration of members and vehicles are the duties of the secretary. Unlike all other office staff, who are usually long-served station workers that have toiled their way up to the top of the branch hierarchy, the secretary is often a lateral entrant and the only one with a background in a white-collar occupation. His recruitment is the result of the low levels of literacy of the other branch members. Whilst his remuneration is usually above the average of the other officers, he is kept strictly away from the funds. As one of Neoplan's chairmen explained to me: 'The secretary has not the trustiness, but he has the pen'; meaning that he is the only one who is able to read and write properly.

The yard staff includes all those who run the transport on the ground. Positioned at the top of the yard hierarchy is the station master (also referred to as 'overseer'), who acts as the principle coordinator and supervisor of the branch's yard staff, and who simultaneously serves as the main intermediary between the yard and the office. Below the rank of the station master are the transport officer (or 'T.O.') and the 'chief driver'. The T.O. is responsible for overseeing roadworthiness and maintenance of the branch's fleet of vehicles. In earlier times, the chief driver used to serve as spokesman-elect of all the drivers of a given branch. Today, however, the role has become mostly obsolete. As few drivers are actually registered members of a branch, with many switching their association at the discretion of opportunity, the chief driver now monitors operations rather than serving as their advocate.

The remaining positions among the yard staff are grouped into smaller subunits, or 'gangs'. They carry out the loading of vehicles, which means 'hustling' for passengers. The number of gangs depends on the number of destinations served by the branch as well as on the amount of vehicles that are preparing for departure at the same time. The basic composition of a yard staff gang includes the 'bookman' (formerly also referred to as 'collector') who is in charge of dispensing tickets for the vehicle, and his group of 'loading boys', who assist him by 'loading' passengers and luggage. The collection of fares falls mainly within the remit of the bookmen, so most, if not all of the income generated by a branch passes through their hands. Standing at the centre of a structurally volatile system of redistribution, their middlemen-like status is routinely scorned both by passengers and other station workers.[14]

The basic distribution of the positions of the office and yard staff dates to the colonial era, when it was adapted from the organisational pattern of the

14 Because of their strategic function in the organisation of the parks (and thus of the transport sector at large), bookmen were repeatedly targeted by governmental interventions. Two weeks after the 1979 coup, for example, Rawlings' junta declared the bookmen 'unproductive and a nuisance to public interest' and banned them from operations with immediate effect (Daily Graphic 1979a). This directive stood in line with the junta's larger and rigorously enforced policy to 'eliminate all middlemen' (Baynham 1985). Yet as it unleashed great unrest among Ghana's station workers, who threatened to go on strike, it was revoked just four days after its implementation (Daily Graphic 1979b).

FIGURE 2.6–2.8 *(left to right) At the secretary's office, 'overseer' overseeing the yard, loading boy at work*
SOURCE: M. STASIK, 2012

Gold Coast railway station personnel (in turn derived from the model of 'the Victorian railway'; see Lacy 1967). A central, and in this regard, most apparent remnant of this legacy is the frequently used designation of all members of the yard staff as 'porters'. Reminiscent of what Clifford Geertz describes as processes of 'involution' (1963), over the course of time, these initial structures have come to be amended by a multitude of additional positions. In Geertz' example of Javanese farmers, it is the combination of abundant labour and scarce (yet remarkably fertile) land that makes for 'introversive tendencies', 'absorbing increased numbers of cultivators on a unit of cultivated land' (Geertz 1963: 32).[15] This 'static expansion', fuelled by continuous labour intensification, was matched with increasing levels of complexity with regards to tenancy relations, cooperative labour arrangements, technological modes of production and social modes of redistribution – a process Geertz compares to the qualities of 'overornamentation' (Geertz 1963: 82) in late Gothic art.[16]

15 Geertz' argument has brought to the scene a host of criticisms (for an overview, see White 1983), most of which are directed at the lack of evidence his otherwise 'brilliant hypothesis' (Knight 1982, quoted in White 1983: 11) suffers from. I draw here on the analytical potency of the concept, leaving aside the question whether or not it is adequate for explaining agrarian relations in Java.

16 Geertz elaborates on that parallel to late Gothic art by quoting Goldenweiser: 'The basic forms of art have reached finality, the structural features are fixed beyond variation, inventive originality is exhausted. [...] Expansive creativeness having dried up at the source, a special kind of virtuosity take its place, a sort of technical hairsplitting (...).' (Goldenweiser 1936, quoted in Geertz 1963: 81).

A similar picture holds true for Ghana's main lorry parks in general, and for Neoplan in particular. Being a long-standing centre of attraction for day labourers, unskilled workers and roadside venturers of diverse professions, the park 'absorbs' an ever-increasing number of 'cultivators' who work its evermore densely populated land, although Neoplan's additional workforce does not 'plough' land (tenancy relations within the yard, however, grow strikingly intricate). The workers come to ply routes, constantly striving to increase the turnover of cars and passengers by way of organisational fragmentation and an elaboration of established practices – raising the levels of institutional complexity and technical virtuosity with every newly incorporated position.

The creation of such appendixed positions often result from a favour that a senior branch member has rendered to a relative or friend, or from an intricate system in which a yard staff member subcontracts an additional worker, usually as a dispatcher or henchman.[17] In other instances, new positions emerge in order to counterbalance the structural weaknesses of the transport sector at large. Attached to the office staff of most of Neoplan's branches, for example, is a varying number of so-called 'associates' who usually consist of branch members considered too 'senior' (that is, old) for the yard work. 'Exempt' from any duty (as well as from authority and income), their continued presence as branch members effectively serves as a substitute for the lack of a pension scheme for retired transport workers, which in turn depends on the officers' goodwill to grant them something from the daily write-offs of the branch.

The most compelling force behind these involuting processes of growth is the continuously heightened level of competition arising from it. As more and more people attempt to eke out a living by finding (or creating) new niches within the yard economy, their influx necessitates a reworking of existing arrangements for catching up with the very results of the on-going involution. Ressembling the users and producers of Zimbabwe's 'kukiya-kiya economy' that Jeremy Jones writes about, Neoplan's workers are ceaselessly compelled to become 'more multiple, mobile and polyvalent than the adversary' (Jones 2010: 293, citing Detienne and Vernant 1991). As I will show in more detail below, this constant strive for outwitting one's rivals on the ground, and for countering the innovations that these rivals in turn keep on concocting, serves as a main drive for the creation of new and ever-wittier practices, which subsequently leads to the introduction of evermore new positions.

17 The newly created labels of work echo the implied levels of ornateness (and of 'technical hairsplitting'), comprising a broad spectrum of derivations of existing designations, such as 'shift master', 'gang leader', '2IC-T.O.' (that is, 'second-in-command-transport-officer'), 'second porter', 'third porter' and so forth.

Two central examples for newly created niches of yard work that can be considered as a direct outcome of involuting competition are the so-called 'shadows' and 'balabala'. The main, and in fact only, task of shadows consists in sitting in the loading vehicle and pretending to be a passenger to give the impression of a bus almost ready for departure. As 'real' passengers begin to buy tickets and to enter the bus, the shadows begin disembarking from it one after the other. Driven by the competition for passengers between Neoplan's branches, the system of 'shadow passengers' evolved to a degree that a 22-seater bus is filled with up to twelve shadows before the first ticket is sold. The balabala have emerged as a fully unsolicited group of station workers that stands outside the frames of the branch organisation. They offer forms of 'guidance' to outsiders, mainly to passengers.[18] They are internally divided between the 'front door' and the 'back door' balabala. The former wait for departing passengers at the station's entrances, offering to escort them to the bus they want to travel on while carrying the luggage for a small fee. The latter scout for arriving passengers, trying to persuade them to carry their luggage to the next step of their journey.

The balabala are excluded from any official share in the income generated from ticket sales. The modes of payment that all branch workers receive are divided between the 'office money' and 'chop money'. The office money stems from an assigned share of all revenues made by the branch's gangs during the course of a day and is disbursed among the office and yard staff at the end of every full shift (usually 24 hours). The latter, particularly all junior gang members, usually come off rather badly. The chop money is derived from parts of the revenue made by each gang from the loading of every single bus (from the sale of bus tickets) and disbursed among those who participated in the loading immediately after its departure.[19] The kind of hand-to-mouth maxim implied

18 The balabala are of the lowest status within the station hierarchies. This is also reflected in the origin of their label: derived from a colloquial expression in Hausa, 'balabala' refers to trivial or deceptive manners of speech and to forms of poorly paid, exploitative physical labour. As one Hausa speaker explained to me: '*Balabala* is when you talk shit or when you carry shit.'

19 One example for the round figure of shares distributed after loading a 22 passenger car to Kumasi, with a ten Ghana Cedi (GhC) fare per seat (thus 220GhC in total) is as follows: the overall 'booking fee' for the branch will be about 50GhC, of which 35GhC is for the office money and 15GhC for the chop money. The remaining 170GhC is the driver's share, which includes all other expenses such as a daily station toll of 1.50GhC for the association (so-called 'welfare money') and the municipal authorities, as well as costs for petrol, road tolls and commission for the car owner. Generally, however, the fees and allotted shares vary significantly between different branches. In fact, even within a single gang they might change multiple times during a day's shift.

FIGURE 2.9–2.10 *On shadowing duty (top), 'front door' balabala (bottom)*

SOURCE FIG. 2.9: M. STASIK, 2012. ORIGINALLY PUBLISHED IN STASIK (2016). REPRODUCED WITH PERMISSION FROM TAYLOR & FRANCIS LTD. (WWW.TANDFONLINE.COM).

SOURCE FIG. 2.10: M. STASIK, 2012.

in the distribution of the chop money, which for many low-ranking yard workers regularly constitutes the only share, adds a further element of fragmentation to the station's involuting competition. Propelled by the progressive influx of additional workforce (or 'cultivators'), the yard staff gangs introduce evermore-refined differentiations between the vehicles they load (categorising them, among others, by age, size, speed and comfort). These internal splinter tendencies – in turn spurred by mutual allegations of embezzlement – come to form the basis for the formation of more and more gangs, of new branches and, ultimately, of new stations.[20]

The process of involution 'spreading outwards' has been significantly stimulated by the general upturn Ghana's economy registered in recent years. This, in turn, has raised levels of monetary circulation, increasing the quantities of travel and stipulating the demands for transportation. Over the last ten years or so, about a dozen of independent transport companies have been established in close proximity to the Neoplan Station, creating a belt of satellite stations – all of which have been founded by former members of Neoplan's branches. Many of these spin-off stations run rather humble services, operating as few as three buses and some two departures daily. Others represent more large-scale enterprises, most prominently the so-called 'VIP Station' located right across the road to Neoplan's yard. Established as a collaborative endeavour between former members of Neoplan's biggest branch and its sister station in Kumasi, VIP operates a growing fleet of luxury buses that ply the route between Accra and Kumasi, competing with Neoplan for travellers that belong to Ghana's aspiring middle-class.

Bringing the Passenger to the Driver

During my fieldwork, the most frequent question that I posed to the station workers (in a variety of ways) was 'what work do you do?' By far the most frequent answer I received (in turn formulated in a variety of ways) was 'we load cars'. Not only was this the most common response I received from bookmen and loading boys, but from all of the many different groups of station workers. Enoch,[21] a seasoned branch worker in the position of a T.O. to whom I served as an assistant for some time, expanded on this answer by posing me a riddle.

20 Within the GPRTU, the formal requirement for the establishment of a new branch is the enlistment of one hundred members. In practice, however, this requirement is regularly circumvented through the manipulation of membership figures (e.g. by registering members multiple times or by including retired or even deceased members).

21 The names in this chapter may or may not have been changed.

'How many people do you need to run a lorry park?' he asked. Before I could collect my thoughts for what I reckoned an appropriately elaborate answer (which would have included 'many'), he unravelled it for me himself: 'Two: one driver and one passenger.' As I asked him what was then the purpose of all the many station workers, including his own, he came up with an answer that summarised the raison d'être of the station in a single sentence: 'We bring the passenger to the driver.'

The station, despite the involuting ubiquity of different engagements evolving with it and within it, is basically about getting cars onto the road. That is, first 'loading' them with passengers and goods, and then dispatching them. The station is a mediating institution; an intersection of roads, which accommodates the diversity of routes that are plied on them as well as the diversity of people that travel on them. The ways in which this mediating function of the station is performed are contingent on a number of variables, including the quantity of routes, cars and travels, the qualities of cars and roads, the distances of the respective routes, as well as the frequencies of travellers.

The departure of vehicles in the Neoplan Station is not organised according to a fixed time schedule as mentioned above, but according to the inflow of passengers. Every vehicle is only allowed to depart when it is 'full', thus when the last available seat has been taken. Referred to as 'fill and run' in the Ghanaian drivers' parlance, this principle prevails in long-distance public transport in many locales across Africa (see e.g. Cissokho 2012: 181; Horta 2013: 81; Peace 1988: 16). The main reason for this flexible mode of departures is economic. With subsidies making up for the loss of 'empty' rides being unavailable, both drivers and station workers – whose income depends solely on the sale of tickets – are urged to fully exploit the given capacity for seats, cargo and mileage.

A direct corollary of the 'fill and run' principle is the system of rostering deployed at Ghana's stations. Founded upon a principle of rotation, it is meant to safeguard an equal share in the revenues generated from the route that each driver and station worker serves and helps to load respectively. The kind of egalitarianism implied in the 'fill and run' principle resonates with a broader socio-economic ethos common across Africa. Ghanaian English trenchantly subsumes this ethos in the proverbial wisdom that 'everybody needs to chop' (read: everybody needs to eat). In the practices of the station workers, the rostering system provides the rule that the first vehicle in the queue of cars for a given route must be fully loaded before passengers are allowed to enter the next vehicle waiting in line; as Ghana's drivers put it: 'first come, first serve'. As demonstrated above, the atmosphere of involuting competition at the Neoplan Station corrupted the principle of rotation insofar as, for about half the routes served from it, there is not one but several (at times even up to two dozen) queues of cars loading simultaneously for the same destination. Overall, however, this subversion of the 'first come' rule affected Neoplan's system of departure only in terms of quantity (and complexity), not in principle.

While the inflow of passengers determines the number and pace of Neoplan's departures, their 'outflow' depends upon the availability of cars. In turn, the availability of cars depends upon the passenger inflow to the destinations that the vehicles ply. When the inflow of passengers at one of the destinations served by a branch is stalled, its cars will become 'stuck' at the other end of that route. And the other way round, when the inflow outpaces the number of available cars (and seats), it is the passengers who will be 'stuck'. This fragile balance is further affected by the distance of the respective routes as well as the conditions of the road and, not least, of the vehicles. If a vehicle gets stuck in traffic, its 'delayed' arrival will also postpone its departure and may subsequently lead to a shortage of available cars; even more so if it breaks down en route or is involved in an accident.

A consequence of the dynamic correlation between the inflows and outflows of passengers and vehicles within the Neoplan Station and its 34 destinations is that the traffic in its yard is constantly oscillating between the two extremes of what can be imagined as a continuum between inordinate levels of congestion and yawning emptiness. At the one end of the continuum are particularly low frequencies of the number of incoming passengers, which leads to a surplus of vehicles waiting for departure in Neoplan's yard. The created spatial congestions foster states of hustle and agitation, stirring up competition and aggression among the branch workers who increase their warring for passengers. At the other end the particularly high frequencies of passengers' overall travels lead to too few cars being available, which in turn stirs up competition and aggression among the passengers who fight over tickets and seats.

Against this general framework in which Neoplan's loading procedures take shape, I now turn towards the loading practices, or as Enoch describes it, the ways in which the station workers 'bring the passenger to the driver'. For this, it is useful to discriminate between non-competitive loading and loading performed in competition. The non-competitive loading is a fairly straightforward affair, and serves well to exemplify the basics of the craft.

Loading to Lagos: The Basics

One example for non-competitive loading is Neoplan's Lagos branch, which is the only of Neoplan's 13 branches that plies just one route (to Lagos). As mentioned above, this long route is served exclusively by small and 'resilient'

FIGURE 2.11 *Passengers 'stuck' due to a shortage of vehicles*
SOURCE: M. STASIK, 2011

saloon cars (mostly five-seater Volvos and BMWs). On a regular day, about five vehicles dispatching 20 passengers make the trip from Neoplan. According to the limited turnover of cars and passengers, the fleet of vehicles registered with the branch is of a manageable size, as is the number of staff. Besides the customary board of five officers, there is only one main yard staff worker who acts as station master, T.O. and bookman: Mohammed.

Mohammed is in his mid-forties and is a so-called 'Agege returnee'. Born in Ghana, he lived for many years in the Lagos suburb of Agege, where he worked as a driver for a Ghanaian-owned company, and later he started an import business for spare car parts. His business went bust and former colleagues encouraged him to return to Ghana and hire out at Neoplan. His background in the very destination he now loads the cars to is a common pattern among Neoplan's branch workers. It mainly follows from pragmatic reasons of knowing the ins and outs of the place and of being able to communicate with the passengers in their respective languages and idioms. Although Mohammed has no permanent group of loading boys, there are usually two youths to assist him in exchange for a daily meal and for the opportunity to sleep in one of the branch's parked cars, which is where he also spends his nights. Being the main and only yard staff responsible for the loading, he is on duty permanently and as such, he has made the station his home.

Because there are no competitors for the Lagos route within Neoplan, Mohammed does not scout for clients, rather, it is the incoming travellers who come looking for him. When a traveller has found their way to him, Mohammed first inquires about the 'immediacy' of their travel as the journey to Lagos is best begun just before dawn, so that all border checkpoints can be passed before nightfall (and before a possible close-down for the night). If the journey is of no particular immediacy, the traveller is offered tea and asked to make him/herself at home – Mohammed's home. Even if the intended trip is urgent, the waiting time for three other passengers to fill up the car will be unavoidable, unless one is willing to pay for the empty seats. The 'urging' passenger will, in any case, be asked to pay an extra fee for what is called a 'hurry car', the sum of which depends on the willingness of the driver and of the other passengers to depart earlier.

As more travellers trickle in, Mohammed begins by assigning them into groups of four according to the sizes of their luggage. Ideally, each group's luggage will fill the capacity of a Volvo trunk. Next, he surveys the travellers' passports, checking for the numbers of entry stamps into Nigeria. Those with so-called 'virgin passports' (passports with no entry stamp at all) pay the highest fare, as a fixed amount is added to the regular fare for 'balancing' the fees (that is, paying the bribe) needed to pass Nigerian customs. This fee reduces with every collected entry stamp. From seven entries upwards, 'virginity' is fully lost and no additional balance levied.

Following these initial inspections of baggage and red tape, Mohammed checks for the order of vehicles to 'go on scale', thus consulting the roster as to which cars are to 'fill and run' next. Then the loading of the luggage starts, which is the moment his assistants are called into action. The process often lasts long into the night, with pieces of luggage repeatedly being rearranged inside a car or exchanged between different cars until a 'perfect fit' is reached. In the meantime, Mohammed negotiates the luggage fees with the passengers which can continue well after all the pieces are stored away inside the trunks. During particularly tenacious struggles over additional luggage fees, Mohammed's favourite (and usually successful) way to leverage reluctant passengers is to order his assistants to unload their respective pieces again, while countering subsequent scoldings with a shrug.

At last, the fares are collected. While the preceding negotiations are usually accompanied by much fanfare, the collection of fares is performed in silence, attesting to the amounts of money paid. Mohammed deducts from the collected amounts the fees for loading – which represent the office money and are subsequently passed on to one of the branch trustees – and for the luggage, which is the chop money. The remaining sum is the driver's share. While the drivers will usually be present during these pre-arrangements, they will not get involved. The loading procedure is the responsibility of the yard staff only, which, in this case, is Mohammed and his boys. Once the passengers are seated and the drivers have received their share, the loading is completed and the responsibility is passed on to the drivers.

Notwithstanding that the practices of loading the Lagos cars need to be worked in order to be sustained, they follow well-developed routines that face little deviations. The most turbulent moments during Mohammed's daily work is when the cars return from Lagos. His role is then to coordinate the rearrangements of the parking order, as the outgoing cars need to make way for the incoming ones and to regroup again within the boundaries of the yard parcel of the Lagos branch. Even with moderate levels of overall congestions on Neoplan's yard, and thus little interference from the traffic of other branches, this re-parking procedure may well take more than one hour, as well as excessive amounts of abusive language traded between Mohammed and the drivers.

During times of heightened demand for transport to Lagos, usually before holidays or when the exchange rate of the Nigerian Naira goes into sudden decline (favouring purchases in Nigeria), Mohammed pre-empts possible shortages of available cars by calling on his extensive network of befriended drivers to be on hold. The driver's share for such a 'spare car' will usually increase significantly (often to twice the amount of the normal share). In most cases this additional cost will not be added to the passengers' fare, but deducted from the office money. While this practice may in fact deprive the branch of any share from the ticket sales (except the luggage fees), it serves to 'keep the market moving', to use one of Mohammed's most frequently uttered dictums. The attitude formulated therein is not one of benevolence, rather, it stems from Mohammed's efforts to make all incoming passengers commit to the cars of his own branch in the long term. For, while there is no direct competitor inside Neoplan, the branch still competes with about ten Accra-based operators that ply the route from other stations. This, however, is about as far as the effects of competition come to impinge on the loading actions in Neoplan's Lagos branch. The general absence of direct competition makes the branch an 'island' of tranquillity in the midst of what otherwise constitutes a 'sea' of tempestuous struggles.

Loading to Kumasi: The Hunt

The roughest scrambles take place among workers loading vehicles to Kumasi. Kumasi is Ghana's second biggest city and hosts what is believed to be West

Africa's largest open air market (Clark 1994). The 370 km long connection from Accra to Kumasi represents the single most frequented route in the country and Neoplan's main destination. Even during the least busy days, the overall number of Kumasi travellers dispatched from Neoplan will normally reach around 2,000. During particularly busy days, this number regularly increases tenfold. Some basic patterns of travel density can be discerned, such as business travellers' propensity for weekday's travels and the tendency of many Accra-based Akan to visit their upcountry folks during the weekend. However, the route is generally subjected to unpredictable fluctuations, which are mainly due to Kumasi's role as a central transit node for connections further north.

The passengers for the Accra-Kumasi route are competed for by four of Neoplan's branches.[22] Two of these branches are in what, in a broader sense, might be termed a 'friendly relation'. Their chairmen established a memorandum of understanding (MoU) whereby one branch will only register (and thus load) vehicles with a carrying capacity of 35 seats and above, whilst the other will limit its loading to cars below 35 seats. The practical effects of this bilateral agreement are fairly limited, however, as the respective cars still need to be 'filled', regardless of their size. Further yet, as the two other branches are not party to this deal, they register and load their cars at will and whichever size. Passengers' own preferences do carry some influence on the type and condition of the vehicles, with novelty and speed being the most important criteria (see also Klaeger 2014: 204–219).[23] Consequently, the most recent (and fastest) car models are usually first introduced on the Kumasi-bound route.[24] Yet due to the diversity of travellers and to the sheer magnitude of their travels, many other, slower and (much) older types are also 'put on scale'.[25] During periods of high density travel, when all four branches approach full capacity, virtually any kind of vehicle available will be deployed in order to handle the number of

22 In addition to that, about five other satellite stations ply the route to Kumasi from locations in close proximity to Neoplan's yard; most prominently, the VIP station.

23 Another common criterion is 'on-board entertainment', usually consisting of one or several flat-screen monitors on which the latest collections of Ghanaian and Nigerian movies are shown for the passengers' diversion. While some newer vehicles come with ready-made screens and audio equipment, in most other cars the screens are installed by local fitters who weld them into a rack mounted at the back of the driver's cabin or on the inside of the roof.

24 In recent years, these comprised primarily of brand-new coach buses of Chinese (Yutong) and South Korean (Daewoo) manufactures, as well as American (Ford), Japanese (Toyota) and South Korean (Hyundai) minivans.

25 By far the most popular model among these older types is the Benz 'Sprinter': a transporter turned passenger van by means of local craftsmanship, including newly installed windows and 23 to 27 seats (depending on the length of the chassis).

incoming passengers. This regularly includes private saloon cars, taxis and at times even motorbikes.

Many more hands are at work on the Kumasi-bound route than in the tranquil loading procedures of the Lagos branch. Whilst the basics of the craft remain the same (getting the vehicle on scale and passengers and their luggage on board), the division of the work and of the tasks increases substantially, due mainly to the dynamics of involuting competition, which here gain full momentum. The impetus behind this accelerated involution comes down to the fact that, rather than simply wait for the passengers to come to them, the yard staff of the Kumasi branches need to actively entice them and get them into their car before their competitors do. The loading of Neoplan's Kumasi-cars resembles a hunt, and although the most common term that the workers use in metaphoric self-designation is 'warrior' (*asafo*) rather than 'hunter' (*ɔbɔfoɔ*), many of their practices resemble those needed for hunting. The following vignette exemplifies some of the main manifestations of these 'hunting' practices. It stems primarily from my own yard work experience at the very 'forefront' of the hunt, which, for most parts, I acquired under the guidance (and command) of station master Rasford.

Rasford's Adzaa-tricks

Rasford works for the 'Odzinga Ford Highway Express' branch. It has the highest turnover of cars and passengers among the four Kumasi branches. The chairman did not sign the MoU regarding the seating capacities of registered vehicles and its relations with the other branches are generally anything but friendly. The branch came into existence about 15 years ago, when a gang of yard staff (led by the then-bookman by the name of Odzinga) decided to splinter from their former branch. As the story goes, these men were 'the fastest loaders of the fastest cars' (Ford minivans) and they decided that they would be better off fending for themselves. The assertive demeanour conveyed in the foundation story of the branch continues to remain present among its workers still today. For Rasford, who was hired by the branch five years ago, the story serves as a kind of vocational ethos. He told me on various occasions that his aim is to 'produce the fastest loaders' of all. Aged 27, he is remarkably young for the position he holds and his rapid advancement within the branch hierarchy is a result of the adept loading skills he demonstrated as a loading boy and, later on, as a bookman. And while many of his subordinate workers are senior to him in terms of age, they approach him with a mixture of admiration and awe.

During each 24 hour shift, which starts at seven a.m. every other day, there are between six and eight (at times, ten) gangs that load Kumasi-cars for the Odzinga branch. Their overall number depends on the inflow of passengers and the availability of cars. Rasford supervises half of these gangs, sharing the responsibility with another station master. Each gang comprises at least nine members, including one bookman, about three loading boys and five shadows. When fewer cars are on scale simultaneously (because of a limited inflow of passengers or of a limited availability of cars), the two station masters merge the gangs, with bookmen taking shifts as loading boys and loading boys positioned as additional shadows. When the number of cars loaded at once exceeds the number of gangs (usually just before peak times of travel), it is the loading boys who come to step in as bookmen, while shadows assist as loading boys. Within these fluctuating arrangements, there are in fact recurring constellations that foster collaboration among the yard staff of different gangs, though more often than not the gangs tend to work against each other. This highly competitive mode is further spurred by the competition with the gangs of the other three branches loading for Kumasi. From the perspective of each gang, a crucial factor that determines the intensity of the competition – and of the ingenuity needed to defy its ramifications – is the location of the vehicle they load within the yard. The further inside the yard the car is parked, the more other Kumasi-cars will the incoming passengers pass by before reaching it, thus making the loading task more difficult. The cars loaded under Rasford's supervision are usually positioned right in the centre of Neoplan's yard. In order to get the travellers into his cars, Rasford instructs his gangs by means of what he calls 'a well-designed psychology'; a 'psychology' aimed at reading the passengers' minds. As he explained to me, it consists of three main '*adzaa*-tricks'.[26]

> The first trick is to rush. You catch the passenger at the entrance, fast. You tell them 'Master, only two seats left'. And you rush them to the car. Like that, you give the feeling of hurry; like they need to hurry to catch the last seat. You make them sweat on the way, 'hurry-hurry-hurry'. And when you reach the car, you seat them right inside. This is manipulation by speed. Then we got manipulation by promise. Like when sometimes they reach the car and they see that it's not filled up, not as they expected. Then you do the trick to make them stay. You promise, 'Ten minutes only. Master, I promise, ten minutes we go. If not, you get refund.' So they agree. You check the ticket, you get the balance, you go. Once they pay, there is no turning back. Even if they call on you for refund, you never refund them. If they quarrel, we show them what's written on the backside: 'Ticket not

26 *Adzaa* is a Hausa term for 'conman' or 'trickster'. It is frequently used by both Neoplan's yard workers and its passengers.

> refundable'. [...] Then we got 'AC-10. This is the trick for *nkurasefoɔ* [village people]. They pay for the rural-type of car, but they like the city-type of treat. When you see them coming in, you tell them now we got the AC [air-conditioned car] for 10 Ghana [Cedi]. Of course this is false. We know AC starts from 15. But they don't have the sense. This is enough to carry them across and make them sit. If they ask the driver for the AC, he just laughs at them. He tells them it goes automated, that it will start when they get into more heat in the bush. But this is not your concern anymore. They check the ticket, they are through.

Rasford is no more reckless in resorting to these 'tricks' than the staff of other gangs loading in competition. In the general 'hunt' for Kumasi passengers, pretence, persuasion and cunning ways of cheating are all long-established and integral parts of the trade. What distinguishes Rasford from most of his colleagues and competitors is the level and promptness of inventions by which he refines the loading skills of his gangs. It was not without pride that he pointed out to me that yard workers of other branches copied the 'manipulations' he claimed 'authorship' of. One invention he was particularly proud of related to the ways in which he advanced the performative register of shadow passengers.

Before his invention, the only 'disguise' shadows used for their work was sitting in the loading bus. Their outward appearance was not different from that of loading boys. And as shadows are regularly drawn on for helping out in the loading, they often come to sit in the bus dirty and soaked with sweat. Unsatisfied with this easily debunked appearance, Rasford began working out ways of adding more credence to the shadow performance. At first, he ordered his shadows to bring along clean spare shirts, which they had to change into when on shadowing duty. He furthermore equipped them with sets of 'props', such as travel bags and briefcases they would place on their laps as well as brochures and books that they would (pretend to) read. The masquerade triggered aggressive reactions on the side of competing branches (all of which deploy shadows as well). During periods of particularly low travel density, which are the times when shadows are most in demand, they formed mobile squads to stalk Rasford's shadows and sabotage their loading by warning potential passengers about the sophisticated bluff. Rather than cutting back on the practices of disguise, Rasford responded by diversifying the shadow 'cast'. He included female hawkers and sellers of various ages to sit in his cars. In return for their service, they were granted privileged access for selling to passengers that boarded subsequently loaded cars. As the unmasking became increasingly difficult (at times leading to awkward encounters in which actual female passengers were accused of shadowing), the competitors abandoned their sabotaging endeavours and instead began making use of Rasford's invention themselves. The latest refinement Rasford advanced the masquerade with to date came to include myself as, what he dubbed, the 'international shadow'.

Obviously, the passengers are not particularly pleased with the many manipulative 'tricks' that are played on them. Upon entering Neoplan's yard, they virtually turn into 'fair game' that is preyed on by the station hunters. Their reactions bespeak of the initial bewilderment, ranging from instances of panic to efforts of self-defence through to masterfully enacted forms of ignoring the surrounding hustle. These kinds of defending responses regularly turn into offensive rants once the passengers realise that the car they were just seated in (and paid the fare for) is not as cheap, new, fast, well-equipped or 'filled' as expected – and that they were thus 'trapped'. About the most common cause for irritation and for subsequent fury is the discovery of shadows, despite the many refinements of the latters' bluffs. The reactions to the fraud often become even more furious when passengers feel betrayed on a more personal level. As cars may take up to several hours to fill before departure, shadows are regularly engaged in long, and intimate, conversations with real passengers. Whilst on shadowing duty, it was more than once that I found myself narrating to fellow passengers fantastic stories about the many marvellous things I planned to do upon reaching the destination that I was never actually going to travel to. Yet for the more veteran passengers travelling via Neoplan, the masquerade is an indication of the finesse and efficacy of the gangs' loading skills. During the course of a longer conversation I had with a market woman who travelled to Kumasi, she 'unmasked' me, telling me that she was well aware of me being a shadow. Having figured out the parameters and ploys of loading, she deliberately picked the car I helped to shadow. If they are able to hire an *oburoni* (white man), she explained, they will 'delay me the least'.

Conclusion

Accra's Neoplan Station stands as a marked representative of Ghanaian lorry parks. Yet this does not imply that all of Ghana's lorry parks work in the same way. Even if we only consider the major parks in cities and larger towns, the spectrum of variations is extensive. The most apparent differences lie in parameters of size, layout and location, which include, for example, makeshift loading bays arranged along a main thoroughfare (e.g. the 'old station' in

Wenchi, Brong Ahafo Region), a chronically congested entrepôt-like yard that is annexed to a large regional wholesale market (e.g. Techiman's central lorry park), a dusty football field (e.g. the station in Sefwi Wiawso, a district capital in the Western region) or the decaying structures of a once-modernist transport terminal transformed into an inner-city bazaar (e.g. Accra's Tema Station). Similarly diverse are the numbers of routes served from a given station, ranging from one principle route plied by cars from the park in a remotely located secondary city to the fifty-plus destinations served from Ghana's biggest road-travel hub, the Kejetia Station in Kumasi. The qualities of registered cars span a broad range as well. For instance, in Accra's Agbogbloshie Station, whose primary clientele are market women trading in vegetables at the adjacent markets, most vehicles qualify as 'boneshakers' (a local expression for old and dilapidated passenger lorries). In Tema's so-called 'Accra Hurry Station', which is mainly used by shipping agents who commute between the harbour and the capital, most vehicles on scale are no older than three years.[27]

Many of these differences are attributable to a combination of historical, geographical and economic circumstances under which the operations of the respective parks take shape. Another factor relates to what might broadly speaking be described as the 'cultural' traits of the park's main clientele. Among Neoplan's yard workers, they are frequently framed in terms of ethnicity. For example, the reluctance of many Ashanti to board worn-out or dirty vehicles is commonly explained by referring to their reputedly noble self-image. As describe to me by a (Ewe) driver: 'The Ashanti have too much pride to ride the goat-bus. They will not entertain the long waiting for the goat-cargo to be loaded. And they will not enjoy the smell for sure.' During my time at Neoplan, I regularly witnessed goats being loaded into Kumasi-cars in which many Ashanti were seated (who usually seemed anything but delighted about the living cargo). The point, however, lies on the side of the (often preconceived) opinions of the station operators, rather than with 'ethnic' markers of passengers' travel standards as such.

If we conceive of these many differences as a kind of 'phenotypic' variations of the Ghanaian lorry park, then one of the central traits inscribed into the 'genotype' is involution. Contingent on a range of external parameters (relating primarily to historical, geographical and economic contexts) there are also significant differences between the degrees of 'static expansion' reached in terms of institutional fragmentation, technical elaboration and 'in-yard' competition.

27 This is not the actual age, but refers to the year that the car was licenced in Ghana (evidenced by the last two figures on the licence plate).

As a general rule of sorts, we might infer that the more lucrative and, in this regard, competitive the routes plied from a given station are, the more apparent will be the 'ornateness' of internal divisions of work and the more honed the virtuosity of loading practices and of 'tricks' involved. The actual ramifications of yard involution, however, appear to follow quite diverse trajectories.

Shadow passengers, for example, are commonly deployed in the major stations in Accra and Kumasi where routes are served in competition. Yet the much-refined shadowing scripts described above are a unique 'virtuosity' of the practices among gangs loading only in Neoplan. In fact, few other lorry parks appear to match Neoplan's complex levels of 'overornamentation' (Geertz 1963: 82). Still, the tendency of involuting growth is detectable across the whole 'phenotypic' range of Ghana's main lorry parks. A main indicator of involuting structures is the competitive loading for the same route of at least two different branches within a single station, which – by default – leads to a multiplication of workforce and thus static expansion. Regarding this 'overdriving of an established form' (Geertz 1963: 82), namely, the branch organisation, Neoplan constitutes no exception whatsoever.

This 'late Gothic' quality of Ghana's public road transport systems has long since incurred disapproval on the side of the authorities as well as the (road travelling) public; not least because it is seen as an untoward sign of the deficient capacities of the state to adequately provide for and to regulate public transport services. Successive governments – of both colonial and post-independence regimes – sought to thwart the processes of involuting growth by way of regulatory interventions. Depending upon the ideology and the larger economic and political constellations of the time, these interventions materialised as restraints imposed via sanctioning forms of law, taxation, certification, price control, import policy, the competition of subsidised parastatals, as well as brute attempts to clamp down on what, in habitual recurrence, was framed as the 'chaotic' operations inside Ghana's main lorry parks.

By and large, the variously configured mechanisms of regulation, intervention and coercion turned out to be vain attempts and, at most, merely decelerated the pace of organisational fragmentation.[28] Similar to other roadside 'trades'

28 The greatest inroads regarding legislative measures were made by the administrations under Rawlings, particularly under the Provisional National Defence Council (1981–1993). Rawlings gave extensive patronage to the GPRTU, strengthening its position in relation to rival associations while counting on the political support of its members. For the time being, this system of clientelism proved largely successful. As noted by Gyimah-Boadi (1994: 132), the GPRTU became a 'prosperous organ' of Rawlings' revolution.

characterised by largely informal and decentralised arrangements – above all, the often ill-famed trade of hawking (see e.g. Asiedu and Agyei-Mensah 2008; Hansen 2004) –, regulation by dint of bylaws, penalties and episodic 'sanitation' campaigns bears little more than placebo effects suited to soothe public opinion before elections. Enoch, the above-mentioned T.O. from Neoplan and my station mentor, summarised the station's rigid position in a long history of regulative vagaries as follows: 'Politics and money can trick the people, but they can never trick the station. The people are many, but the station is more.' In other words, being greater than the sum of its many and much dissociated parts, the station's increasing fragmentation becomes a mechanism to immunise it against attempts to curb or to monopolise its involuting operations. At the same time, it is the inherent trait of involution that facilitates an evermore expanding density and diversity of road travels despatched from its yard.

References

Ademowo, Adeyemi J. 2010. '"They are Poor and Violent": Stereotype and the Ibadan Urban Motor Park Space.' *Antrocom* 6 (2): 115–133.

Albert, Isaac O. 2007. 'Between the State and Transport Unions: NURTW and Politics of Managing Public Motor Parks in Ibadan and Lagos, Nigeria.' In *Governer les villes d'Afrique: État, Gouvernement Local et Acteurs Privés*, edited by Laurent Fourchard, 125–139. Paris: Karthala.

Asiedu, Alex B. and Samuel Agyei-Mensah. 2008. 'Traders on the Run: Activities of Street Vendors in the Accra Metropolitan Area, Ghana.' *Norwegian Journal of Geography* 62 (3): 191–202.

Awedoba, Albert K. 1981. 'A Note on Lorry Names in Ghana.' *Journal of the Anthropological Society of Oxford* 12 (1): 63–64.

Barrett, Jane. 2003. 'Organizing in the Informal Economy: A Case Study of the Minibus Taxi Industry in South Africa.' *SEED Working Paper* 39. Geneva: ILO.

Bähre, Erik. 2014. 'A Trickle-up Economy: Mutuality, Freedom and Violence in Cape Town's Taxi Associations.' *Africa* 84 (4): 576–594.

Bakare-Yusuf, Bibi and Jeremy Weate. 2005. 'Ojuelegba: The Sacred Profanities of a West African Crossroad.' In *Urbanization and African Cultures*, edited by Toyin Falola and Steven S. Salm, 323–340. Durham: Carolina Academic Press.

Baynham, Simon. 1985. 'Divide et Impera: Civilian Control of the Military in Ghana's Second and Third Republics.' *Journal of Modern African Studies* 23 (4): 623–642.

Beisel, Uli and Tillmann Schneider. 2012. 'Provincialising Waste: The Transformation of Ambulance Car 7/83-2 to Tro-tro Dr. JESUS.' *Environment and Planning D: Society and Space* 30 (4): 639–654.

Boeke, Julius H. 1953. *Economics and Economic Policy of Dual Societies*. New York: Institute of Pacific Relations.

Cissokho, Sidy. 2012. 'Réformer en situation de "décharge": Les transports publics à Dakar durant les mandats de Abdoulaye Wade.' *Politique Africaine* 126: 163–184.

Clark, Gracia. 1994. *Onions are My Husband: Survival and Accumulation by West African Market Women*. Chicago: University of Chicago Press.

Clayborne, Dontraneil D. 2012. *Owner-Drivers in the Tro-tro Industry: A Look at Jitney Service Provision in Accra, Ghana*. PhD thesis, University of California.

Coulibaly, Moussa. 1993. *Les Transport Urbains au Senegal de 1945 à 1970: Le Cas de Dakar*. MA thesis, Université Cheikh Anta Diop de Dakar.

Czeglédy, André P. 2004. 'Getting Around Town: Transportation and the Built Environment in Post-Apartheid South Africa.' *City & Society* 16 (2): 63–92.

Daily Graphic (Ghana). 1979a. 'Bookmen Banned.' July 19.

Daily Graphic (Ghana). 1979b. 'Bookmen at Lorry Parks Not Banned.' July 23.

Daniel, Ikuomola A. 2012. 'Poverty, Road Culture and Deviance among Areaboys in Lagos Central Neighbourhood.' *International Journal of Prevention and Treatment* 1 (1): 1–10.

Date-Bah, Eugenia. 1980. 'Inscriptions on the Vehicles of Ghanaian Commercial Drivers: A Sociological Analysis.' *Journal of Modern African Studies* 18 (3): 525–531.

Davies, Justine. 2008. 'Selling Wares on the Streets of Accra: A Case Study of Street Hawkers in Ghana's Capital.' *Focus on Geography* 51 (3): 32–36.

Detienne, Marcel and Jean-Pierre Vernant. 1991. *Cunning Intelligence in Greek Culture and Society*. Chicago: University of Chicago Press.

Feld, Steven. 2012. *Jazz Cosmopolitanism in Accra: Five Musical Years in Ghana*. Durham: Duke University Press.

Field, Margaret J. 1960. *Search for Security: An Ethno-psychiatric Study of Rural Ghana*. New York: W.W. Norton.

Fouracre, P.R., E.A. Kwakye, J.N. Okyere and D.T. Silcock. 1994. 'Public Transport in Ghanaian Cities: A Case of Union Power.' *Transport Reviews* 14 (1): 45–61.

Geertz, Clifford. 1963. *Agricultural Involution: The Process of Agricultural Change in Indonesia*. Berkeley: University of California Press.

Goldenweiser, Alexander. 1936. 'Loose Ends of Theory on the Individual, Pattern, and Involution in Primitive Society.' In *Essays in Anthropology Presented to A.L. Kroeber in Celebration of His Sixtieth Birthday*, edited by R. Lowie, 99–104. Berkeley: University of California Press.

Grieco, Margaret; Nana A. Apt and Jeff Turner (eds). 1996. *At Christmas and on Rainy Days: Transport, Travel and the Female Traders of Accra*. Aldershot: Avebury.

Guseh, James S. 2008. 'Slogans and Mottos on Commercial Vehicles: A Reflection of Liberian Philosophy and Culture.' *Journal of African Cultural Studies* 20 (2): 159–171.

Gyimah-Boadi, Emanuel. 1994. 'Associational Life, Civil Society, and Democratization in Ghana.' In *Civil Society and the State in Africa*, edited by John W. Harbeson, Donald Rothchild and Naomi Chazan, 125–148. Boulder: Rienner.

Hansen, Karen T. 2004. 'Who Rules the Streets? The Politics of Vending Space in Lusaka.' In *Reconsidering Informality: Perspectives from Urban Africa*, edited by Karen T. Hansen and Mariken Vaa, 62–79. Uppsala: Nordic Africa Institute.

Hansen, Karen T. and Mariken Vaa (eds). 2004. *Reconsidering Informality: Perspectives from Urban Africa*. Uppsala: Nordiska Afrikainstitutet.

Hart, Jennifer. 2011. *'Suffer to Gain': Citizenship, Accumulation, and Motor Transportation in Late-colonial and Postcolonial Ghana*. PhD thesis, Indiana University.

Hart, Jennifer. 2014. 'Motor Transportation, Trade Unionism, and the Culture of Work in Colonial Ghana.' *International Review of Social History* 59 (22): 185–209.

Hart, Keith. 1970. 'Small-Scale Entrepreneurs in Ghana and Development Planning.' *Journal of Development Studies* 6 (4): 104–120.

Hart, Keith. 1973. 'Informal Income Opportunities and Urban Employment in Ghana.' *Journal of Modern African Studies* 11 (1): 61–89.

Hawkins, Edward K. 1958. *Road Transport in Nigeria: A Study of an African Enterprise*. London: Oxford UP.

Hill, Polly. 1962. 'By Air to Niamey.' Working paper, University of Ghana, Legon.

Hill, Polly. 1963a. *Statistics of Ownership of Mammy Lorries in Ghana*. Legon: Institute of African Studies, University of Ghana.

Hill, Polly. 1963b. *The Migrant Cocoa Farmers of Southern Ghana: A Study in Rural Capitalism*, Cambridge: Cambridge University Press.

Hill, Polly. 1965. 'Notes on the Kumasi Cattle and Sheep (and Goat) Markets.' Working paper, University of Ghana, Legon.

Hill, Polly. 1984. *Indigenous Trade and Market Places in Ghana, 1962–64*. Jos: University of Jos.

Horta, Gerard. 2013. 'Interurban Collective Transport and Road Crashes in Santiago, Cape Verde.' *Etnográfica* 17 (1): 77–95.

IBIS Transport Consultants Ltd. 2005. *Study of Urban Public Transport Conditions in Accra, Ghana*. World Bank and PPIAF.

Jalloh, Alusine. 1998. 'The Fula and the Motor Transport Business in Freetown, Sierra Leone.' *African Economic History* 26: 63–81.

Jennische, Ulrik. 2012. 'Traders, Drivers and the National Health Insurance Scheme in Small Town Ghana.' *Urban Forum* 23 (4): 467–481.

Jones, Jeremy L. 2010. '"Nothing is Straight in Zimbabwe": The Rise of the Kukiya-Kiya Economy 2000–2008.' *Journal of Southern African Studies* 36 (2): 285–299.

Jordan, J.D. 1983. 'Public Transport in Harare.' *Zambezia* 11 (2): 127–138.

Jordan, James W. 1978. 'Role Segregation for Fun and Profit: The Daily Behavior of the West African Lorry Driver.' *Africa* 48 (1): 30–46.

Joshi, Anuradha and Joseph Ayee. 2002. 'Taxing for the State? Politics, Revenue and the Informal Sector in Ghana.' *IDS Bulletin* 33.

Khayesi, Meleckidzedeck. 2002. 'The Struggle for Regulatory and Economic Sphere of Influence in the *Matatu* Means of Transport in Kenya: A Stakeholder Analysis.' *Proceedings of the Sixth International Conference on Competition and Ownership in Land Passenger Transport*, Cape Town, South Africa, Sep 19–23, 1999.

Khosa, Meshack M. 1992. 'Routes, Ranks and Rebels: Feuding in the Taxi Revolution.' *Journal of Southern African Studies* 18 (1): 232–251.

Klaeger, Gabriel. 2009. 'Religion on the Road: The Spiritual Experience of Road Travel in Ghana.' In *The Speed of Change: Motor Vehicles and People in Africa, 1890–2000*, edited by Jan-Bart Gewald, Sabine Luning and Klaas van Walraven, 212–231. Leiden: Brill.

Klaeger, Gabriel. 2012. 'Movements into Emotions: Kinetic Tactics, Commotion and Conviviality Among Traffic Vendors in Accra.' In *Urban Life-Worlds in Motion: African Perspectives*, edited by Hans-Peter Hahn and Kristin Kastner, 131–156. Bielefeld: Transcript.

Klaeger, Gabriel. 2014. *Speed Matters: An Ethnography of a Ghanaian Highway, its Perils and Potentialities*. PhD thesis, University of London.

Knight, G. Roger. 1982. 'Capitalism and Commodity Production in Java.' In *Capitalism and Colonial Production*, edited by Hamza Alavi, 119–159. London: Croom Helm.

Kwakye, E.A. and P.R. Fouracre. 1998. *Urban Transport Policy Reform in Ghana*. Berkshire: Transport Research Laboratory.

Kyei, Kojo Gyinaye and Hannah Schreckenbach. 1975. *No Time to Die: A Book of Poems by Kojo Gyinaye Kyei, Depicting Slogans of Ghana's Mammy Lorries with Photographs and Illustrations by Hannah Schreckenbach*. Accra: Catholic Press.

Lacy, Leslie A. 1967. *A History of Railway Unionism in Ghana*. MA thesis, University of Ghana, Legon.

Lamote, Frederik. 2007. *De vloeiende stad: trans(nationaal) urbanisme: verhalen over Gare Petersen, Dakar*. MA thesis, K.U. Leuven.

Lawuyi, Olatunde B. 1988. 'The World of the Yoruba Taxi Driver: An Interpretive Approach to Vehicle Slogans.' *Africa* 58 (1): 1–13.

Lewis, Barbara. 1970. 'Ethnicity, Occupational Specialisation, and Interest Groups: The Transporters' Association of the Ivory Coast.' *African Urban Notes* 5 (3): 95–115.

Lewis, George H. 1998. 'The Philosophy of the Street in Ghana: Mammy wagons and Their Mottoes. A Research Agenda.' *Journal of Popular Culture* 32 (1): 165–169.

Lopes, Carlos M. 2009. 'Hug Me, Hold Me Tight! The Evolution of Passenger Transport in Luanda and Huambo (Angola), 1975–2000.' In *The Speed of Change: Motor Vehicles and People in Africa, 1890–2000*, edited by Jan-Bart Gewald, Sabine Luning and Klaas van Walraven, 107–126. Leiden: Brill.

Mashiri, Pedzisai. 2001. 'Managing "Face" in Urban Public Transport: Polite Request Strategies in Commuter Omnibus Discourse in Harare.' *Zambezia* 28 (1): 85–95.

[MRH and SSGH] Ministry of Roads and Highways and Statistical Service Ghana. 2009. *National Transport Household Survey Report*. Republic of Ghana.

Mutongi, Kenda. 2006. 'Thugs or Entrepreneurs? Perceptions of Matatu Operators in Nairobi, 1970 to the Present.' *Africa* 76 (4): 549–568.

Ndiaye, Mamadou and Rémy Tremblay. 2009. 'Le transport routier au Sénégal: Problématique de la gestion des gares routières.' *Canadian Journal of Regional Science* 32 (3): 495–210.

Ntewusu, Samuel A. 2012. *Settling In and Holding On: A Socio-Economic History of Northern Traders and Transporters in Accra's Tudu: 1908–2008*. Leiden: African Studies Center.

Nugent, Paul. 2010. 'Do Nations Have Stomachs? Food, Drink and Imagined Community in Africa.' *Africa Spectrum* 45 (3): 87–113.

Okpara, Enoch E. 1988. 'The Role of Touts in Passenger Transport in Nigeria.' *The Journal of Modern African Studies* 26 (2): 327–335.

Olubomehin, Oladipo O. 2012. 'The Nigerian Motor Transporters since the 1920s.' *International Journal of Humanities and Social Science* 2 (12): 230–237.

Peace, Adrian. 1988. 'The Politics of Transporting.' *Africa* 58 (1): 14–28.

Pritchett, Jack. 1979. 'Nigerian Truck Art.' *African Arts* 12 (2): 27–31.

Quayson, Ato. 2010a. 'Signs of the Times: Discourse Ecologies and Street Life on Oxford St., Accra.' *City & Society* 22 (1): 72–96.

Quayson, Ato. 2010b. 'Kòbòlò Poetics: Urban Transcripts and their Reading Publics in Africa.' *New Literary History* 41 (2): 413–438.

Quayson, Ato. 2014. *Oxford Street: City Life and the Itineraries of Transnationalism*. Durham: Duke University Press.

Seck, Serigne M. 2006. 'Transport et territoires: Les coxeurs de Dakar, du bénévolatà la prestation de service.' *Bulletin de la Société Géographique de Liège* 48: 7–17.

Sesay, S.M. 1966. 'Drivers in the Transport Industry: A Case of Road Transport in Sierra Leone.' *Sierra Leone Studies* 19: 86–97.

Simone, Abdoumaliq. 2004. 'People as Infrastructure: Intersecting Fragments in Johannesburg.' *Public Culture* 16 (3): 407–429.

Spittler, Gerd. 2001. 'Teilnehmende Beobachtung als Dichte Teilnahme.' *Zeitschrift für Ethnologie* 126 (1): 1–25.

Stasik, Michael. 2015. 'Rhythm, Resonance and Kinaesthetic Enskilment in a Ghanaian Bus Station.' *Ethnos* (online first).

Stasik, Michael. 2016. 'Contingent Constellations: African Urban Complexity seen through the Workings of a Ghanaian Bus Station.' *Social Dynamics* 42 (1): 122–142.

Stasik, Michael and Alena Thiel. 2014. 'Market Men and Station Women: Changing Significations of Gendered Space in Accra, Ghana.' *Working Papers of the DFG Priority Programme 1448* (No. 5), Leipzig and Halle.

Stoller, Paul. 1989. *The Taste of Ethnographic Things: The Senses in Anthropology*. Philadelphia: University of Pennsylvania Press.

Tambulasi, Richard and Happy Kayuni. 2008. 'Can the State Perpetuate the Marginalisation of the Poor? The Socio-economic Effects of the State's Ban on Minibus "Callboys" in Malawi.' *Development Southern Africa* 25 (2): 215–226.

Titus, S.I., E.D. Andrew and K.C.S. Mynepalli. 2010. 'Refuse Disposal Practices in Three Major Motor Parks in Ibadan Municipality, Nigeria.' *Journal of Public Health and Epidemiology* 2 (4): 82–86.

Trans-Africa Consortium. 2010. *Report on Statistical Indicators of Public Transport Performance in Africa*. Brussels: Union Internationale des Transports Publics.

Turner, Jeff. 1996. 'Trading and the Tro-Tro: A Description of the Informal Public Transport Sector of Accra.' In *At Christmas and On Rainy Days: Transport, Travel and the Female Traders of Accra*, edited by Margaret Grieco, Nana A. Apt and Jeff Turner, 19–33. Avebury: Aldershot.

van der Geest, Sjaak. 2009. '"Anyway!" Lorry Inscriptions in Ghana.' In *The Speed of Change: Motor Vehicles and People in Africa, 1890–2000*, edited by Jan-Bart Gewald, Sabine Luning and Klaas van Walraven, 253–293. Leiden: Brill.

van Walraven, Klaas. 2009. 'Vehicle of Sedition: The Role of Transport Workers in Sawaba's Rebellion in Niger, 1954–1966.' In *The Speed of Change: Motor Vehicles and People in Africa, 1890–2000*, edited by Jan-Bart Gewald, Sabine Luning and Klaas van Walraven, 75–103. Leiden: Brill.

Verrips, Jojada and Birgit Meyer. 2000. 'Kwaku's Car: The Struggles and Stories of a Ghanaian Long-Distance Taxi-Driver.' In *Car Cultures*, edited by Daniel Miller, 153–184. Oxford: Berg Publishers.

wa Mungai, Mbugua and David A. Samper. 2006. '"No Mercy, No Remorse": Personal Experience Narratives about Public Passenger Transportation in Nairobi, Kenya.' *Africa Today* 52 (3): 51–81.

Weiss, Brad. 2009. *Street Dreams and Hip Hop Barbershops: Global Fantasy in Urban Tanzania*. Bloomington: Indiana UP.

White, Benjamin. 1983. '"Agricultural Involution" and its Critiques: Twenty Years After Clifford Geertz.' *Working Paper Series of the Institute of Social Studies* (No. 6), The Hague.

Wrangham, Elizabeth 2004. 'An African Road Revolution: The Gold Coast in the Period of the Great War.' *Journal of Imperial and Commonwealth History* 32 (1): 1–18.

CHAPTER 3

Jam-space and Jam-time

Traffic in Nairobi

Amiel Bize[1]

In a Jam

I am sitting in a *matatu*, a minivan bus, on Uhuru Highway entering the central business district of Nairobi. It is Saturday afternoon, I am coming from work on Mombasa Road, and the roads are packed – workers leaving jobs in Nairobi's industrial area, people running Saturday errands, everyone's moving at the same time. The jam is on. Grandly decorated, larger 45-seater matatus booming loud dancehall music, 60-seater buses in more staid corporate branding, hulking trucks marked 'transit goods only', and taxis are jammed in with a swarm of cars – battered old cars, luxury sedans, tinted and chromed SUVs and the loud-engines and decals of a growing street-racer culture. Motorcycle delivery bikes drive through the lines of stalled cars. We breathe the exhaust. Around us is the hum of engines, chatter on phones and then the loud revving of engines and honking as space opens up and the ribbon extends and contracts, dragging us forward. As Keguro Macharia writes, 'in Nairobi's now-infamous jams, one encounters time as sound – a temporal choir emerging from stalled cars' (Macharia 2010).

I'm up front, with the driver and a chatty out-of-town man who assembles and sells exercise equipment. This guy is talking a mile a minute, but the driver looks like he is about to fall asleep. I try to chat with the driver, to keep him awake; I make the obvious comment about the jam, I ask about his working hours, how much sleep he gets. Very little, maybe four or so hours a night. He naps during the day and when he is driving the adrenaline keeps him awake, but he says that sleepiness overcomes him when he is stuck in the jam. The heat, the exhaust, the waiting. If we passengers are trapped in the kind of time that is neither work nor leisure, what Lefebvre (2002) calls 'constrained time', the driver has a more serious dilemma – for him, time *must* be productive.

1 I'm grateful for the comments and corrections provided by Kurt Beck, Basil Ibrahim, Gabriel Klaeger and Michael Stasik, as well as the assistance of Alfred Anangwe. The PhD research on which this article is based was supported by a Wenner-Gren Foundation Dissertation Research Grant, a National Science Foundation Dissertation Improvement Grant and a Social Science Research Council International Dissertation Research Fellowship.

 | DOI 10.1163/9789004339040_004

As we approach the roundabout where the matatu will turn off Uhuru Highway and go down Haile Selassie Avenue toward the Railways bus terminus, we can see the new traffic lights. These digital traffic lights began to be installed in July and August 2013;[2] now, in September, people sometimes follow them and sometimes do not. At this particular time, as always during rush hour, the traffic police are ignoring the lights and directing traffic through the roundabout. Alone in their rhythm, the lights count down absurdly long amounts of time, using letters and numbers: D-10, 9, 8, 7… then C, then B, then A, *then* down from 100. Watching them is as infuriating as waiting for the traffic police to wave you forward, but the contrasting pacing momentarily provides amusement and a topic of conversation, an opportunity to mock the traffic police who, as the driver tells me, cannot keep up with the new 'digital' era.[3] Nairobi Governor Evans Kidero claimed that traffic police would be removed from roundabouts on March 1st, 2014, but, like many other state-imposed deadlines, the day came and went with no change.

Finally we near the roundabout and I ask the driver if it is alright to get out into traffic, not wanting to get him in trouble with the police or the much-feared City Council *askaris* (guards; also called *kanjo*). He seems surprised I would want to walk when we are finally almost there, but I am eager to escape. The gym equipment guy also gets out and goes in a different direction and so, like the cars released from the jam in Cortázar's (1973) story 'The Southern Thruway', where friendships formed in a years-long traffic jam suddenly dissipate when the cars begin to move, our brief shared moment is over.

Jam-space and Jam-time

In 2010, the African Development Bank (2010) projected that Nairobi's population growth rate to 2025 was, at 77.3%, the second highest in Africa. Although much of Nairobi's new population is unable to access motorised transport, nevertheless this urban growth results in a city that has, according to an IBM study, ten times the number of residents that its road network can support. Unsurprisingly, then, Nairobi ranked fourth-worst among twenty cities surveyed by IBM for its 2010 'Commuter Pain Index' (IBM 2010). The study surveyed commuters' impressions of traffic and travel time and the effects of these on travel

2 Nairobi has had traffic lights on and off for many years; they come in and out of use. These 'digital' traffic lights are therefore just the newest iteration (and expenditure).

3 During the most recent elections, Uhuru Kenyatta (now president) and his running mate William Ruto branded themselves as 'digital' and the opposition as 'analogue'. This entered everyday conversation through jokes and comments as reflecting a broad binary of progress and backwardness.

decisions; Nairobi commuters reported among the worst delays due to traffic, with a mean delay of two hours. Traffic congestion seems unlikely to lessen in the near future, as many of the underlying factors – land use and settlement patterns, urban population growth, credit liberalisation and transport policies encouraging automobility – have not been addressed. Jams thus figure large in the texture of everyday life in Nairobi, claiming huge amounts of time for commuters and, I will suggest, even for the large percentage of the urban population that cannot afford to ride a bus or matatu, much less own a car. They are also a rhetorical vehicle through which the vicissitudes of Kenyan society and urban life are discussed, so people not only spend time *in* jams, they also spend time talking *about* them. As both an everyday experience and a subject of conversation, jams shape ideas about time, about work, about progress and about governance.

Beyond Nairobi, traffic jams play an important role in discourse around the rapid urbanisation taking place across the African continent. If the expansion of automobility holds out the promise of development, because roads and automobiles are both metaphorically and practically vehicles of capital circulation, traffic jams represent its dark side. Iconic images of clogged, chaotic traffic have come to stand for the larger problems of African cities, revealing their failure to properly 'manage' their growth. Wasting time, fuel and money, jams are understood as profoundly unproductive spaces – the image of cars 'idling' encodes both the destructive waste of exhaust fumes and the waste of potentially productive work time.

Nairobi's new governor, Evans Kidero, has made battling traffic jams a key part of his larger goal of making the city a safe, clean, business-friendly capital. Picking up on the discourse about jams as unproductive spaces, he has linked this to a new concern: 'The city has to be improved to attract more investors' (Mwangi 2014). The circulation of vehicles is explicitly associated with the circulation of capital, not only in a logistical sense but also a representational one – traffic 'chaos' gives the city a bad image and dissuades investors. Waste and chaos thus serve parallel discursive functions, adding a moral dimension to the jam; Kidero's response is to 'tame' traffic by focusing on behaviour and as such to use traffic as a tool of moral regulation. Such discursive use of traffic jams as barriers to efficiency and productivity occludes, however, their more complex role as urban spaces and times. Jams are populated by a range of figures who seek to eke out bits of value from these interstitial spaces; they make time difficult to control; they can be claimed as territory and even turf.

As a contribution to a volume on the 'making' of African roads, this chapter argues for the need to recognise the centrality of *work* in making roads what they are. Historically, mobility in the city has been linked to the patterns and needs of people going to work. Kenda Mutongi (2006) has shown that the matatu industry grew up as a response to the transport needs of Nairobi's new population who, following independence, migrated into the city in large numbers and had to be able to travel from the new 'African' neighbourhoods to work in other parts of the city. As the city has grown, many workers continue to commute from their homes to workplaces. But the road is also a work*site*, both for drivers who move along the road and for a diverse set of merchants who mark out space for themselves on and alongside specific roads. The tight connections between Nairobi's 'formal' and 'informal' worlds are thus visible in traffic jams, both spatially and temporally – not only does the space bring together commuters, city officials, transport workers and mobile vendors, but the different spatial and temporal configurations of salaried and unsalaried work also intersect in the jam. The rush hour jams created by the travel patterns of salaried workers create interstitial space-time for commuters and a mobile and part-time workplace for unsalaried workers. Jams thus both shape and exemplify the non-standardised temporalities of unsalaried work.[4] In focusing on the importance of work in 'making' roads, then, I also argue for the importance of time and temporality as analytics for understanding the road.

A second argument concerns the discourse of waste within which jams are most commonly framed. As Vinay Gidwani argues in *Capital, Interrupted*, conversations about waste both reveal theories of value and make governance possible by justifying the interventions of those who speak them (2008:19). In the case of jams, government rhetoric proposes productivity and efficiency as the 'values' counterposed against the jam's waste. It is certainly true that being stuck in traffic jams, particularly when 'time is money', is exhausting for transport workers and commuters alike. But alongside this dichotomy of efficiency and waste, the jam reveals and makes possible other possibilities and subjectivities. This chapter shows the ways that the jam destabilises clock time, produces class subjectivities, allows for extractive governance and creates a site for 'marginal gains' (Guyer 2004). (These possibilities or subjectivities do not necessarily escape the logics of control and exploitation, but they work against a simple understanding of jams as 'wasteful'.) I begin with a discussion

4 This chapter does not have space to address the insufficiency of the concepts of 'formal' and 'informal' for describing the life and political economy of the many ways of creating livelihoods in Nairobi, many of which depend on participation in multiple spheres of activity. Though I occasionally use these terms, I find that discussions of precarity and flexible labour – which are, in their way, quite formal – might be more suited to a discussion of unsalaried workers in Nairobi.

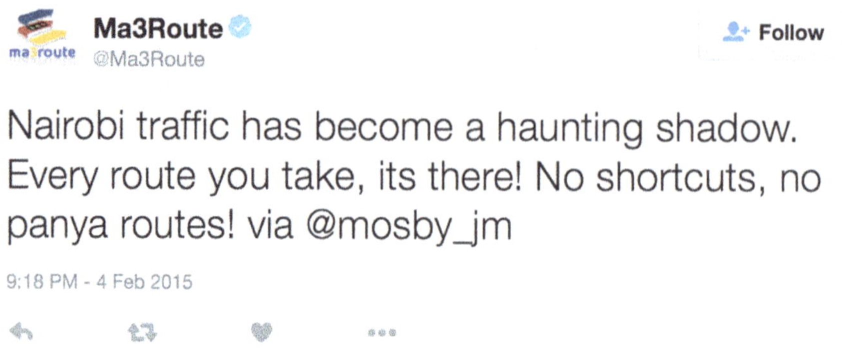

FIGURE 3.1 *A tweet about morning traffic: 'No shortcuts, no back routes!'*
SOURCE: TWITTER.COM, 2015

of the historical conditions under which a rapid expansion of automobility has become possible, desired and to some extent necessary and then consider how jams operate in the imagination and everyday lives of a range of Nairobi residents.

Automobility and Urban Kenya

The number of vehicles on Kenyan roads doubled between 2001 and 2009, across most vehicle classes.[5] This growth reflects historical, economic and social dynamics that have made automobility both necessary and desirable. A rapid rate of urbanisation; an emphasis on road expansion; classed perceptions of automobile ownership; the importance of vehicles as an investment or business strategy; and expanded access to both imports and personal credit have encouraged this rise in vehicle numbers. Such a rapid expansion of automobility also suggests that many of the troubling dynamics of Kenya's growing urbanisation will be maintained, even entrenched, in the near future: segregated settlement patterns, urban congestion and air quality concerns and oil dependency (Klopp 2011). For instance, a 2013 article suggested that the country's oil imports were to increase by a third over the following two years,

5 According to the Kenya National Bureau of Statistics. Both private cars and passenger vehicles doubled. Though other vehicle classes grew at a slower rate, the overall number doubled because of an 8-fold increase in the number of motorbikes. A policy to open up the passenger motorbike sector as an employment opportunity for young men made motorbikes tax free in 2008, sparking enormous growth in this area (http://www.knbs.or.ke/REGISTERED%20 VEHICLES.php).

due largely to the growth in the country's 'auto fleet' (*Reuters*, 28 November 2013). By the end of 2014, oil comprised 33% of Kenya's total import bill (Ngigi 2015).

Unsurprisingly, the increasing number of vehicles has led to what Mark Lamont calls 'infrastructural overload'. As Lamont puts it, citing Brenda Chalfin, Africa is experiencing a consumer *perestroika* and 'the speed at which transport infrastructure is being constructed lags well behind the sheer numbers of automobiles being imported' (Lamont 2013: 381). Traffic congestion is one of the major signs of this overload, and it is used to justify all manner of policies and statements about the problems of urban life. In terms of transport, congestion problems are mobilised especially in the service of a transportation policy that focuses on road building – just the kind of policy that is likely to encourage more cars, but one which has become common sense for Nairobi residents and politicians alike. Though it is a truism among transportation scholars that building roads is a short-term fix that ultimately does not solve traffic congestion problems, historical dynamics have meant that roads are among the most important of 'public goods' – thus, building roads is a political act and one that is much celebrated (Manji 2014).

The emphasis on good roads and smooth traffic has important historical antecedents, having to do with the racialized dynamics of transportation and mobility in colonial times, with the importance of automobility as a sign of modernity and self-reliance at the time of Kenya's independence, and with a longstanding political economic system of private gain from public investment. Infrastructure played a role in debates over whether Kenya was ready for self-rule (Slaughter 2004) and first president Jomo Kenyatta placed great emphasis on kilometers of asphalt as a barometer of development and freedom from dependence.[6] At a more local level, roads were seen as public goods to be distributed, and under both Kenyatta and his successors, Daniel arap Moi and Mwai Kibaki, politicians earned credit by 'giving' roads to their constituencies, even while the fortunes they milked from their role in distributing road-building contracts often resulted in lower road quality. The perception that Moi-era road building was particularly bad – profit skimming leading to surfaces that were 'thin' and quickly wore off – made 'good roads' a major part of third president Mwai Kibaki's attempt to build his legacy. Among other projects, Kenya's Vision 2030 plan to become a middle income country includes road building as a central element of infrastructural policy.

6 Relatedly, residents of marginalised regions also use kilometers of paved road (only 100 km in the North Eastern Province, for example) to demonstrate their marginalisation.

In a 2011 article on transportation policy in Kenya, Jacqueline Klopp nicely summarises the history of segregated land use and fragmented transportation policy that has led, directly or indirectly, to the current expansion of automobility and marginalisation of non-motorised forms of transport. Restrictions on movement and settlement under colonisation meant that African residents were initially excluded from formal planning in Nairobi. This created a separation between the low-density, Northwestern suburbs where Europeans lived and the informal high-density African settlements in the Southeastern parts of the city. As Klopp writes, segregated living created a dependence on personalised transport among Europeans, enforced distances between living and working spaces for labourers and was the beginning of an urbanisation strategy in which informal strategies fill in the gaps of formal planning. This segregated land use pattern is still visible in Nairobi, though racial segregation has become social segregation (Klopp 2011). The 'leafy' suburbs and gated communities that house wealthier Kenyans (and expats) remain primarily accessible by car, and those who do walk – mostly people working in the residences – find little dedicated pedestrian infrastructure, forcing them to walk not just on the edge but *in* the road.

Much of the land outside the city has gradually been given over to housing and real estate and since the early 2000s newer settlements like Pipeline and Mlolongo along Mombasa Road have begun to grow exponentially. These are largely unplanned highrise developments that fill a need for low-cost housing. Higher-cost housing has also been built outside the city, with projects like the Thika 'superhighway' and the Northern and Southern Bypasses fueling the fantasy of traffic-free access to the city for a new middle class commuter population. Former Maasai and Akamba grazing land outside the city, in Ongata Rongai, Kitengela, Syokimau and beyond, is also being converted into housing for a middle class that increasingly drives. Residents of these areas, which Nairobians now call the 'diaspora', acknowledge the trade-off between being able to afford to build on their own plot of land (an important aspiration) and the increased commute time. Those living in the 'diaspora' find adaptive ways to avoid, or manage the jam and its time implications – staying in town after work, taking night classes, catching up on sleep in public transport or leaving home at five a.m. to miss rush hour. If mobility was an important part of work in the colonial period when workers moved between labour pools in the 'tribal reserves' and urban worksites, it remains extremely important, both for commuters and itinerant labourers like repairmen whose work depends on their movement through the city.

Transport policy, much like housing, has historically underserved the needs of lower income residents. Responding to the rapid urbanisation that took place after independence, when restrictions on urban residence were loosened, Kenya's first president Jomo Kenyatta deregulated the passenger transport sector, allowing matatus to take the place of the formal bus services that had proved inadequate. According to Kenda Mutongi (2006), Kenyatta saw matatu operators, who mostly operated illegally, as embodying the spirit of 'entrepreneurship' that was necessary to develop the new nation. He celebrated both their entrepreneurialism and the fact that they offered a vital service in an urbanising economy by providing a means for people living in informal settlements on the outskirts of Nairobi to travel to the industrial and clerical centres. This semi-institutionalisation of what was initially an *ad hoc* solution to both colonial and postcolonial inability to provide adequate transport has created a form of public transport that is regulated in part by interests beyond the needs of its users: the profit structure relies on the ability of larger collectives to incorporate smaller-scale owners, making control over routes extremely important. Matatus provide far better service than public transport options in the majority of U.S. cities, but they do have their problems: they are associated with violence (wa Mungai and Samper 2006), liable to deviations in route and hikes in fare and often unaffordable for the lowest-income residents.

Today, mode of transportation reflects social status more accurately than terms like upper, middle and lower class – there is a class of Kenyans who are driven, those who drive themselves, those who take public transport and those who walk.[7] Given the fact that more than half of Nairobi residents are pedestrians, while the bulk of infrastructure is targeted toward drivers, it seems clear that current policy perpetuates the exclusion of a large portion of Nairobi's population. This policy, Klopp writes, emphasises automobility 'at the expense of gender, equity, sustainability and poverty concerns' (2011: 3).[8]

This, too, is visible in the jam. On the cover of a recent report on inequality in East Africa, a photograph shows a young boy peering into the window of a Porsche stuck in traffic. This image comes with a set of intuitions about what is going on within it – presumably the boy is asking for money from the cars'

7 As with any categorisation, one can find even finer distinctions: on certain routes, there are more and less expensive buses, or buses for younger versus older people; there are people who walk to work but could take a bus in bad weather or if they are late etc.

8 It also excludes environmental concerns. Road development and city sprawl are damaging existing ways of reducing the jam's pollution. The construction of new suburbs and roads on Nairobi's eastern boundary, for instance, has led to the excision of 60 acres of the Nairobi National Park, which is reputed to act as a cleaning system for Nairobi's filthy air.

passengers, who are invisible behind tinted windows. The exposure of the boy and protection of those within reveals one form of inequality, but the fact that such an expensive car is stuck in traffic speaks to a dynamic in which all urban residents are subject to the jam. The image also suggests that automotive policy has come to shape urban life in ways that go beyond its impacts on mobility, because the road is not just for driving on. Even for those without cars, roads are public spaces and sites of work. In the remainder of this chapter, I consider how cars and traffic jams shape subjectivities and possibilities for work on and along roads.

Rush Hour

Nairobi's daily jams respond to and make visible the ebb and flow of working life in the city; not only the formal labour of those who commute but also the range of informal activities that spring up to 'service' the jam. While not all jams are rush hour jams, it is the everyday jam that reveals most about urban patterns. More and more, 'rush hour' threatens to take over the whole day; roads are clogged for more daytime hours than they are clear. This exceptional time is increasingly the norm.[9] This occasions intense frustration, as expressed

FIGURE 3.2 *Rush hour traffic on Mombasa Road*
SOURCE: A. BIZE, 2014

9 As Rudolf Mrázek writes of Jakarta, it is clear streets that are exceptional. He links traffic to the ability to read events taking place in the city. 'Streets were jammed = there was no riot. Streets were empty = there was a riot somewhere' (2004: 433). In Nairobi, similarly, empty streets suggest that an event of potential violence is taking place.

in a retweet from the account @Ma3Route on February 22, 2015: 'How can like half this city be gridlocked on a Sunday afternoon?!? #Nairobi is doomed! How long till 100% gridlock?'

The Kenyan government now claims that traffic jams cost 50 million shillings a day in 'lost productivity' (McGregor and Doya 2014),[10] and Nairobi's Governor has made battling traffic jams a key element in his claim to technocratic governance. In January 2014, half a year after taking office, Kidero launched a Transport and Urban Decongestion Committee. Citing the need to make the city more attractive to investors and to reduce the waste of time and fuel that traffic jams incur, the committee was to 'formulate policy on attitude and behaviour change among drivers, touts, passengers and pedestrians' (Mwangi 2014). While the focus on attitude and behaviour is not the most obvious of strategies for decongestion (though it is common in road *safety* campaigns), it reveals something of the moral landscape within which traffic jams are couched. Because of a longstanding association between driving behaviour and social responsibility more generally – one which focuses especially on 'rush' and 'reckless' driving as an expression of social ills – even traffic jams come to be addressed within a behavioural framework. Responding more recently to criticisms of his failure to address jams, Kidero tweeted: 'I propose that people should go to work very early in the morning, like me.'

Seeing traffic order as social order is a trope of urban thought (Conley 2012, Goffman 1971). But in imagining a jam full of office workers on their way to work or meetings, discussions of 'lost productivity' reveal the extent to which Nairobi residents are imagined as salaried workers, even as the reality is quite different. Taking up this vision of the jam and complicating it, this section considers rush hour's most iconic participants: commuters and the matatu crews who provide them with transportation. Here I am concerned not with any particular jam but rather with 'the jam' – a repeating feature of Nairobi life that shapes, as problem of governance and as daily experience, the subjectivity of urban residents. Seen as an urban problem, jams become the vehicle for moralising rhetoric that targets a range of actors, especially matatu drivers. This rhetoric, which brings together a discourse against waste and idleness (or, in this case, idling) with an emphasis on time management and self-discipline, is familiar to us as one aspect of a shift from pre-industrial to industrialised forms of work (Coetzee 1988; Thompson 1967). But the experience of submitting oneself to the jam on a daily basis suggests that traffic also conditions people in ways that challenge the categories of time and space commonly associated

10 This number is lower than an earlier figure of 28 billion shillings annually; the article does not reveal where the number comes from.

with industrial modernity, especially the depiction of time as a scarce resource to be 'managed' and 'used', and the spatial-temporal separation between work and leisure.

Rush Hour Driving: When Time is a Scarce Resource

The term 'rush hour' was coined in England in the late 1880s to describe the most congested part of the day, when commuters going to work or returning home crowded means of transport. In Nairobi, the idea of rush hour also picks up on the accusatory charge associated with the term 'rush'. Being in a rush indicates that one has failed to manage one's time well, and traffic jams are seen as both a cause and a consequence of this state of being. This means that attempts to manage rush hour traffic jams propose to solve the problem of 'being in a rush' (when traffic jams delay people and make them impatient) even as they imply that if people were not in such a rush, there would be no jam to begin with (suggesting that the way people move when they are in a hurry is what causes the jam). 'Rush' thus has a doubled function: it both recognises a contemporary condition in which time is scarce, and serves as an accusation.[11] The doubled sense in which 'rush' operates means that it appears frequently as a trope in the larger idiom of disapproval that characterises traffic talk.

Disapproval is most frequently directed at matatu drivers and their lack of 'discipline' – this is what is most commonly said to cause traffic 'mayhem' (the term mayhem displacing the fact that jams are, in their way, often quite orderly). As a *Business Daily* article on rush hour jams entering the Nairobi central business district reports, passengers approve of (even encourage) matatus' wild driving – weaving around stuck cars, driving on the pavement, on the wrong side of the road – because they are in a hurry to get to work. 'Each passenger is desperately hoping that nothing untoward will happen. The delicate balance between the dangerous drive and the time saved is their ticket to getting to work on time' (Mbugua 2011; see also Lamont 2013). Although this article explicitly ties drivers' behaviour to the needs of the passengers, most discussions of matatu driving see matatus as the *source* of chaos and rush. Indeed, matatus have been banned from Nairobi's central business district (only the larger, more formalised buses are allowed in), even as parking for private vehicles is being increased.[12]

---

11 The critique of rush is also embedded in the Swahili saying *pole pole ndio mwendo,* or 'slow/gentle is the way to go'. *Pole pole* refers both to pace, meaning slow, and behaviour, meaning polite or gentle.

12 This is a particularly salient example of the way that the government mobilises frustration with jams and the 'motor-ridden' city to secure support for its measures, even when

To understand the particular state of being that 'rush' entails, it is worth remembering how time came to be something to be managed by revisiting the industrial-era shifts that E.P. Thompson describes in his seminal essay 'Time, Work-Discipline, and Industrial Capitalism'. Time, Thompson suggests, was not always something that could be saved or lost (1967: 59), but as the irregular work patterns of artisan and even agricultural life gave way to more regularised wage labour in industrialised societies, a moral complex around 'wasted time' emerged. Mechanised industry did not invent time scarcity – even the irregularity of artisan labour included moments where work was heavy and time was short – but the new 'disciplined' time entailed the idea that time was something that could and must be controlled. Significantly, recognising the importance of controlling time did not always result in regularity, as the increased significance of and reliance on clock-time opened doors to its manipulation. Thompson picks up here on Marx's insight that the conversion of time to money is not a fixed rate but rather a variable one: once time can be calculated as money, it can also be manipulated. In the factories, management might change the clocks to lengthen the working day and shorten the dinner hour. Thus, the emerging struggle about work was fought as a battle over time: if the first generation of factory workers were 'taught by their masters the importance of time', by the third generation they were striking for overtime. Thompson writes, 'they had accepted the categories of their employers and learned to fight back within them. They had learned their lesson that time is money, only too well' (1967: 86).

Thompson's essay has been repeatedly critiqued and revised (May and Thrift 2001), but disciplinary understandings of the problems of rush hour traffic in Nairobi fit remarkably well within the moral framework around wasted time and idleness that Thompson describes. This goes for both commuters and drivers: even though transport drivers are not wage labourers, their work is shaped by the tempos of salaried work. These shared tempos mean that for drivers and salaried workers alike, the shape of time is defined by the maxim that time is money. But for both (especially for drivers) the jam's irregularities mean that the rate of conversion of time to money is not fixed. Subject to the frustrations of stalled mobility and the uncertainty of time in a traffic jam, drivers and commuters are forced to be idle against their will. The moral framework governing approaches to time is at odds with the reality of everyday life.

At work on the road, Nairobi's matatu drivers live the jam's rhythms in their own doses of adrenaline and boredom. Because their wage is daily – and

---

the policies themselves – which privilege private cars and secure elite interests – are likely to *increase* jams.

because as a result their financial obligations beyond work are also organised daily – a bad jam can have serious consequences. Their time problem is further complicated by their relationship to public authorities who seek to regulate mobility and in some cases earn extra income by intervening in matatu operations. I met a driver at Nairobi traffic police headquarters who had been arrested around noon and was waiting for the vehicle's owner to send him some money so he could pay the fine and be on his way. He was not particularly irked about having been arrested, which he saw as a part of the job. But he was annoyed that it had interrupted his last chance at a good run before the traffic jam started. By one o'clock in the afternoon, he said, the day is lost – the jam sets in. The jam, along with other work conditions, means that time is structured unevenly for matatu workers. After doing a 'squad' (a round trip), a matatu will return to wait in line – between one and three hours – before leaving again. Waiting time alternates with squads which they rush to complete as fast as possible, because they must complete a minimum number every day in order to pay the owner's daily fee and earn their wage, whatever is left over.

Jam talk, particularly the discourse of rush, reveals that Nairobi residents feel compelled by a need to move quickly that is constantly contradicted by their circumstances. Safety campaigns urging drivers to 'rest' suggest that this contradiction and the 'rush' that it produces have very real effects on drivers. I spoke with two driving school instructors who both articulated this state of being and offered, from a pedagogical perspective, their techniques for dealing with it. 'Chief' was the head driving instructor at Rocky Driving School, the largest of the driving schools that have proliferated around Kenya since the 1980s. He began by repeating a familiar refrain – 'people don't manage their time well and end up rushing' – but went on to articulate a more complex position. For Chief, time management was the key to driving overall, but it was something that had to be learned in a context in which time itself is not predictable. Time bends on Kenyan roads: you start on a six-hour trip, he said, but nine or ten hours later you still have not arrived. Driving students thus need to be trained not only in the rules of the road but in the practice of submitting to time, particularly in traffic jams: 'You cannot drive unless you were trained in that jam', he said. 'Whenever you have a jam, you must prepare earlier, you must wake up earlier and you must start your trip earlier… otherwise you will not reach your destination.' Learning to drive in a traffic jam requires a kind of bodily entrainment, one that can only be learned by inhabiting the jam. The jam requires preparation, but offers no guarantees. And if you do not learn to prepare for the jam, you fall prey to 'stress'. Chief was not optimistic, though, that his students could learn patience: 'you know', he said, 'a human being is just a human being'. Another driving school instructor, John, offered a parallel diagnosis of rushing as a social failure incurred by the instrumentalisation of both time and other people. For John, traffic was not merely something to pass through toward a destination – he explained that learning how to drive in this kind of traffic was a process of learning how to relate to others, to appreciate others, something most people do not do. According to him, the problems of the road were tied to the destruction of the social as a result of transactional thinking: 'The mentality of "you've got to give so you can have" is doing damage on the road.'

John suggests that a sense of reciprocity is broken down by a condition of rush that leads one to instrumentalise others. Learning to cope with the jam entailed, for him, both relinquishing control and submitting to the unpredictability of time. This suggests a vision of reciprocity as mutual submission, and a new moral orientation toward time. As an editorial bemoaning traffic jams comments, impatience only makes things worse: 'I have taught myself to be very patient but I see no reciprocity in our people. I firmed up this thinking after it took me three hours to drive a three-kilometre distance' (Ndemo 2013). If a moral argument against both wasted time and rushing becomes visible in discourse around traffic jams, the actual experience of the jam subverts and complicates such morality by disposing urban citizens – particularly the commuting middle classes and those who serve them – to *submit* to the uncertainty of time.

Commuter Subjectivity

Henri Lefebvre (2002: 52), commenting on the increased importance of leisure time in industrialised France and on writers' corresponding interest in leisure as a theoretical category, wrote that France was undergoing a revolution in values; values associated with work were disintegrating and those associated with leisure were taking their place. But, he argued, another category of time was increasing even more rapidly than leisure time. This was *temps contraint*, 'constrained time'.[13] Neither work nor leisure, *temps contraint* is a time of obligations. In Kristin Ross's reading of Lefebvre, these are obligations that, 'like the departmental cocktail party, are neither precisely work nor, in any real sense of the term, pleasure'. 'Nothing', she writes, 'approximates constrained time better than the space-time of commuting' (1996: 21). If the division between work and leisure was already a feature of the 'modern world', forced idleness appears as an insidious undermining of that distinction.

13 This is more commonly translated as 'compulsive time', but I prefer Kristin Ross's translation as 'constrained time'.

The commute has quintessentially illustrated the ways in which salaried work expands beyond the limits of salaried time, and the jam only exacerbates this expansion. The commute is both unpaid and expensive. Nairobi commuters must literally *spend* the extra hours spent caught in traffic, often incurring costs in the form of care or home work that must be done by other people, or paying higher fares when transport prices go up to compensate matatu crews for being stuck in traffic. A friend of mine who lives in the 'diaspora' leaves the house with her husband at five o'clock in the morning to avoid the jam. They go to exercise at his club, then she goes to work. She teaches evening classes so that she can avoid the evening jam and often they don't get home until ten o'clock. She has three children whom she rarely sees awake during the week.

At the same time, being subjected to the jam is one of the ways in which middle class Nairobi residents develop a subjectivity linked to a sense of themselves self as salaried workers. Stories about traffic, and particularly about the escapades of matatu drivers, are a staple of Nairobi conversation among the commuting classes – who, unlike in other parts of the world, are largely bus and matatu passengers rather than drivers. Mbugua wa Mungai and David Samper's (2006) work on passenger stories in Nairobi sees the narrative genre of the 'matatu story' as a means of coping with city life. Though they focus on violence and stress (an important term across Nairobi conversations), the narratives themselves suggest an emergent class dimension to matatu-based mobility. One story describes a moment of solidarity created when a matatu trying to escape traffic hits a Land Rover. Celebrating the ensuing chaos, the passengers join together in mockery of the 'rich man'. In another story, a passenger fantasises about what it would be like to be the president and speed in Nairobi streets, since only the *wabenzi*, the political class named after their Mercedes Benzes, can access urban streets free of traffic (Mungai and Samper 2006). This commuter subjectivity is not only defined against Nairobi's wealthier classes, of course. It is also contrasted with the class that remains *outside* the vehicle, through stories about phones and bags snatched through windows while commuters are stuck in the jam, or about carjackings. There is a particular combination of arrogance and vulnerability in such conversations; a shared sense that to be on the road in Nairobi is also to be exposed to the coming together of divided worlds.

As a time-space both defined by and separate from work, the jam constructs a class subjectivity that is also very urban, defined through and against the messy worlds of Nairobi roads. Certain public transport vehicles have even begun to recognise something like commuter 'style' recently, referencing (American?) commuter culture by decorating their vehicles with signs that say 'wifi on board' or that show icons of bus riders with computers. Many actually have wifi. Alongside the hip hop, basketball and counter-cultural references of most vehicles, these icons suggest a new imaginary of transport – one linked to connectivity and global consumer culture. Indeed, being online is one way that drivers and passengers spend their jam-time. Drivers tweeting from within the jam fill the feeds of @Ma3Route and @NairobiTraffic, transport information services with twitter accounts. And as one of these tweets suggests, drivers stuck in the jam are thinking about what else they could be doing: @durkchild writes, in Nairobi's abbreviated slang, 'Ngong Rd at Prestige [mall] isn't moving, you can leave your car on the road, go have tea in the mall, pick someone up, get them to sleep with you…'.[14]

Some commuters use the jam's irregularity and interstitial aspects to their advantage. The jam makes people late, and people also use it as an excuse for being late. My friend Alfred tells me that he often blames the jam, he says, when he has not 'allowed enough time' for a particular commute (this formulation itself implies that it is his responsibility to account for and absorb the possible extra time of the jam). I regularly hear people making calls from the bus we're both on, claiming the jam as an excuse before the bus has even left, or saying they are stuck in traffic just outside town when, in fact, they are nowhere near town. But his friends, Alfred says, go even further in using the jam as an excuse:

> There is a day I was drinking with a friend and his wife called. My friend told his wife that he was in a terrible traffic jam when in actual fact he was in a bar drinking. When the wife asked why there was a lot of noise in the background, the man replied 'ni muziki wa matatu' [it's the matatu music]. I told him that he was exaggerating his lies and he told me that when he arrives home he will say that the matatu finally ran out of fuel while in the traffic jam so he had to alight and wait for another matatu.

Here, the jam allows Alfred's friend to escape the obligations of home, to extend his away time or to create an area where he alone controls his use of time. The jam is an ideal excuse for this kind of flexibility, as it makes the relationship of time and space unpredictable. If motorised transport is said to have compressed space-time, jams can also extend it, or, more precisely, destabilise it. The next section considers how this interstitial time also 'thickens' space, allowing what would normally be passed-by to develop into a site of dense exchange.

---

14 @Ma3Route, Feb 4, 2015, 'Ngong Rd at Prestige isn't moving, unaeza acha gari kwa barabara ukunywe chai prestige, ukatie mtu akupromise vitu…'

If jams do provide a certain freedom to extend uncategorised time, however, it is important to note that this 'extra' does not come without a cost. The term 'exhaust', that which spews or trickles from idling cars, nicely metaphorises the wearing nature of traffic jams. Jams tire out people, materials and the environment. Breakdown vehicles are found at roundabouts because cars run out of fuel and need to be towed; people fall asleep at the wheel or in a bus; sometimes drivers turn off their car motors to save energy. For those outside the vehicle, exhaust can also refer to the fumes that perform 'slow violence' (Nixon 2011) on their bodies.

Nyayo Roundabout: Selling in the Jam-space

The Nyayo roundabout, situated adjacent to the Nyayo National Stadium in Nairobi, is notorious for its traffic jams and for the mobile hawkers who take advantage of the jam to sell products to captive commuters. Indeed, Nyayo's jams make it the site of a part-time market, one that appears and disappears along with the traffic. This part-time market suggests that the jam allows a new kind of space to emerge, one that is impermanent but nevertheless dense with social interaction and exchange. As Filip De Boeck (2012) writes, interruptions on the road 'thicken' the public by allowing time for interaction among passengers and between the road and the roadside. This suggests that motorised transport, which quintessentially 'compresses' space-time (Harvey 1989) or 'annihilates' the space between the beginning and end points of a journey (Schivelbusch 1987: 33), can also produce the opposite effect. As a surfeit of motors slows mobility, in-between space becomes newly important. This section examines the way that the jam becomes embedded in the infrastructure and socio-economic life of the roundabout, focusing in particular on the hawkers that work in and alongside traffic.

Intersections are always potential sites of jamming, and at Nyayo, several important roads converge. Each has its own character: Mombasa Road is the main artery linking the East African hinterland to the port of Mombasa; Langata Road conveys residents of the southern suburbs and satellite towns into the city proper; and Lusaka Road connects to the low-income residential areas of Eastlands and Nairobi's industrial area. The character of the roads as much as the activity taking place on them makes the roundabout a space of encounter and mixing. Crossroads have been written about as sites of potentially threatening encounters (Bürge, this volume; Shaw 2002), but roundabouts are intended to reduce encounters and streamline the intersection of vehicles. They only work, however, as long as they are not overloaded. And in Nairobi, they are decidedly overloaded. This fact, as well as their association with British colonial planning, means that roundabouts are often described as outdated, and castigated as a *source* of traffic jams. Proposals to reduce jams often include the removal of roundabouts, which is discussed as a gesture of modernisation – a modernisation symbolised by flyovers that allow motorists to soar above the congested city. Thus, in Nairobi, roundabouts are synonymous with jams, and clusters of infrastructural forms aimed at servicing blocked traffic are often found at roundabouts. This cluster includes tow trucks/breakdown vehicles; billboards or banners (to be read by commuters paused in traffic); and petrol stations that allow people to make use of the slowdown, to keep idling cars from dying or simply to get ahead in traffic. These infrastructural features reinforce the association and lend a kind of permanence to the jam – even when the jam is not on, the structures that service it are built in.

Nyayo roundabout features several of these repeating infrastructures, but the most obvious infrastructural presence is the stadium. Nyayo Stadium gives the roundabout its name and reveals some of the political histories that have created and are embedded in public space in Nairobi. It is named for Kenya's second president, Daniel arap Moi, whose slogan was *nyayo* (footsteps), indicating that he was following in the footsteps of Kenya's first president Jomo Kenyatta. These days, *nyayo* is associated with a range of failed populist projects, and their failure indexes the corruption of Moi's regime.[15] However, the fact that there have not been efforts to introduce similar projects by subsequent governments demonstrates the decline of the state's role in the provision of public services and the rise of the privatisation of public space. The Nyayo roundabout, with its stadium now regularly rebranded by different corporations, fits into a layered political history, one that is interestingly reflected in discourse about traffic jams. One respondent commented that traffic jams at roundabouts revealed Moi's disinterest in progress, where progress is understood as much more spectacular flyovers soaring over clogged roundabouts. 'Moi kept us in the dark for many years. He should have done what Mwai Kibaki [Kenya's third president] has done with Thika Road' (that is, build 'superhighways', bypasses and flyovers).

15 *Nyayo* was a slogan Moi took for his regime to placate the powerful faction of the ruling elite that feared it would lose its privilege following the passing of the country's first president, Jomo Kenyatta. As Moi's brand, though, *nyayo* came to refer more to the new president's form of power than any tribute. If *uhuru* (independence) was the name Kenyatta left strewn about on highways, parks and even future presidents (his son Uhuru Kenyatta is Kenya's current president), linking his presidency to the coming to being of the nation itself, *nyayo* refers more specifically to Moi's form of governmentality: both his insinuation of himself into the everyday lives of Kenyan citizens and a swathe of populist state projects attempted by successive Moi governments during the 1980s.

The desire to fly over traffic jams is also a desire to avoid the 'thick' public (De Boeck 2012), the activities that exist in and around jams, bogging people down and reminding road users of the failures of modernity (Mrázek 2002). This vision of the road – in which one gets trapped by the people and services associated with the jam, rather than caught in the jam itself – finds expression in another government proposal, to solve traffic jams by removing hawkers from roundabouts. A car-centric perspective is apparent in this reversal of causation, by which hawkers and roadside kiosks are said to *cause* traffic rather than be drawn by it. For now, though, these interactive uses of road space have yet to be displaced. Nyayo roundabout is packed with activities and encounters of all kinds. The road edges are bus stops, the far sidewalk is a semi-formal market place for second-hand and new clothing and the near sidewalk a market for informal traders selling second-hand shoes at peak periods of pedestrian traffic. The pedestrian footbridge rising over Mombasa Road is essentially an advertising space; its concrete floor and stairways are filled with rubble and empty of pedestrians. And the road itself, when the jam is on, is a moving market.

Interstitial Space-time and the Part-time Market

Writing of Lagos, the Harvard Project On the City describes 'jam-space' as a 'totally negotiable, usually illegal and hugely productive space' (Bélanger et al. 2001). This description of the jam as a place of plenty is perhaps overly optimistic – and Koolhaas's project has been widely critiqued – but it is nevertheless useful for counteracting the more common association of jams and

FIGURE 3.3 *A hawker selling in traffic*
SOURCE: A. BIZE, 2014

waste. Though workers in the traffic jam work under extremely precarious conditions, it is nonetheless a site that attracts hopeful entrepreneurs by the dozens. The exhaust, threat of injury, suspicion from drivers and harassment by city council *askaris* do not deter them from finding space among the cars to earn a living. I never saw more than forty or so at a time, but several people told me that more than a hundred hawkers worked at Nyayo. In the fullness of a traffic jam, the road is full of sellers.

Sellers working in the Nyayo Stadium traffic jam are, like the jam, part-time. Not only do they leave during off-peak hours, when there is no jam, but they may also work only part of the month. Indeed, jam salesmanship requires being attuned to the ebb and flow of urban life. Hawkers must be sensitive to the urban rhythms created by the circulation of money and people in order to find value in interstitial spaces and 'wasted' time.

In this sense, the urban environment produces a kind of 'oecological time' (Evans-Pritchard 1939), a time defined by environmental conditions and the activities linked to them. Hawkers' selling activities respond to the alternating patterns created by the jam's 'seasons', which themselves respond to the environmental conditions created not only by nature but by salary structures, weather and daily or weekly work schedules. Nairobi residents often comment on the fact that the severity of traffic jams waxes and wanes along with monthly salaries. People get paid at the end of the month, so in the following two weeks more people can afford to gas up their cars and drive to work. As the month progresses, however, and money gets tighter, people go back to riding matatus and the number of cars on the road decreases. Hawkers appear and disappear along with jam. Indeed, many hawkers have other jobs and during the times of the month when pockets are light, they may absent themselves from the jam more often. While it is common in Nairobi for people to have multiple strategies for earning money – a way of life encompassed under the general term 'hustling' – traffic jam hawking makes this more obvious: even the worksite itself is constantly morphing, present at some times and not others.

The jam's transience in space and time also conditions what and how hawkers sell. Goods sold must be light enough to be carried as they walk along the rows of idling cars or chase after a car that has started to move while a transaction is taking place. They must be inexpensive enough to allow quick decision-making by buyers and to ensure that losses brought on by City Council raids are not unendurable. But within these guidelines, the possibilities are great. At Nyayo alone, I have seen hawkers selling fruits and vegetables, phone credit, hats, toys, blankets, cushions, shelves, pens, nuts, stuffed animals, etch-a-sketch pads, paintings, leather bags, jackets and towels.

Some of the Nyayo roundabout hawkers reported selling different items in morning traffic and in evening traffic. Paul, for instance, sells newspapers in the morning and snacks in the evening. Morning commuters, he says, have had their breakfasts and are not interested in food; they are interested instead in finding out what has happened nationally and internationally. People buy snacks in the evening, he says, because they have been too busy to eat during the day or because they want to bring something home to their children. The desire to bring something home for the kids dovetails with the relatively recent availability of low-priced Chinese commodities to offer a proliferation of plastic toys. In fact, several sellers described their goods as items buyers do not know they want or need until they see them, or, intriguingly, as the kind of things people do not budget for (socks, handkerchiefs, portraits of the president). The interstitial nature of the jam is reflected in the form of value that can be derived from it: things people did not know they needed, one person's 'extras' (a snack, a newspaper) that add up to another's livelihood.

Hawkers learn to shape their strategies to suit the conditions of passing automobiles. In his work on roadside sellers in Ghana, Gabriel Klaeger (2012) writes that they become 'engaged' by the features of the 'moving workplace' (Klaeger 2012: 538). Sellers learn to attune themselves to the rhythms of the road and adapt themselves even to the psychological conditions of travellers. By creating a sense of rush, for instance, they can make travellers waiting to depart feel that something is happening and momentarily dispel the boredom of waiting (Klaeger 2012: 549). Klaeger's work shows that automobility shapes the embodied responses of those who work on the road as much as those travelling along it, but also suggests that hawkers provide an affective service in addition to the goods they sell.

Sellers' techniques of persuasion also adapted themselves to the rhythm of the automobile. They were adept at a rapid assessment of whether a given driver or passenger was a likely customer, based on the car, the age of the occupants, perceptions of class and even on gauging drivers' moods. As sellers walk through the corridors of idling cars, their eyes move back and forth, scanning the faces of the people inside the cars, only occasionally bending down to look in the windows. Several vendors I talked to mentioned looking at the faces of people in the cars, particularly at their eyes, intercepting the gaze between eyes and product. This link between eyes and goods opens a channel that the hawkers follow with a sales pitch.

A few hawkers mentioned their *own* facial expressions. One seller, only 15 years old, said he tries to look deserving of pity (playing on the class dynamics materialised in the metal barrier dividing those inside the car from those without); another, a young woman, said she smiles at men, many of whom buy because they want to flirt with her. As Klaeger notes, kind of affective labour seems to be taking place alongside the exchange of goods, and one that is manifest not only in the slippage between sales transaction and date, or sales transaction and charity. Several sellers saw themselves as offering a kind of entertainment. By 'talking nicely' to drivers, they claimed, they were making connections even if the person did not buy. I was surprised to learn that drivers sometimes give money to the sellers without buying, or give more money than what was agreed, even after haggling. The repetitive nature of commuting allows relationships of patronage to form even in what appears to be an in-between space; in this way, space becomes place.

Though sellers, like matatu drivers, calculate earnings daily, they also have monthly obligations that remain defined by the rhythms of salaried labour. These obligations are shaped in part by the expectations of rural kin, who perceive city work as salaried work. One seller, Kariuki, stressed repeatedly that obligations toward his family in the rural area intensified at the end of the month when he was expected to send money. For this reason, he worked harder toward the end of month, spending longer hours in the jam or in other activities. Like E.P. Thompson's early industrial craftsmen, he lived both within the time-structure of wage labour and outside it – though he was bound to a monthly schedule by others' expectations, his work pattern within it was irregular.

Controlling Space

During jam-time, the separation between road and roadside becomes more difficult to maintain. The road is no longer just 'passing through', but becomes a space to be inhabited and worked in. Indeed, the hawkers' presence and the recurrent nature of their contact with one another and with commuters means that the sellers are part of the jam and its meanings. Nevertheless, the ways in which they inhabit the jam are shaped by its flux and impermanence as much as by the repetition that makes the jam a durable space. Furthermore, their claim to a right to work in the jam-space, which relies on the sense that no one has exclusive rights to road space, is challenged by city authorities. These authorities also claim the right to profit from the road's traffic.

Sellers working the jam rely on trust, collaboration and enforcement to manage the mobile nature of their worksite. Jams can 'release' cars unexpectedly and sellers might be caught by surprise. They rely on the customers to pay even if the jam suddenly starts moving; they also rely on other sellers to help them catch up with cars, make change and punish offenders. Though sellers are not organised into formal groupings (like buying groups or savings groups), certain informal agreements are necessary to make their work possible. Since customers need their change quickly, before their vehicle moves forward, a seller might borrow from the person nearest to him with the understanding that accounts will be settled later. Similarly, the fact that they cannot carry all of their goods at once requires secure storage space and relationships with roadside vendors. Some sellers leave extra goods on the ground in the planted median strip (notoriously dangerous spaces in Nairobi), others hang them from trees in the same area, using black plastic bags to store, carry and display goods; others leave them with kiosk owners.

These forms of trust are not left up to chance, though; they are actively maintained, even under the threat of force. Almost like a guild, sellers control the practice of their craft within the Nyayo space, promoting shared norms, sharing information and imposing sanctions on each other. Their self-employed members also impose barriers to entry on new sellers. This is not surprising. Given that the number of migrants to Nairobi from Kenya's rural areas far exceeds the opportunities for formal work, a prime site for salesmanship like Nyayo is subject to being overrun. One hawker, Mary, said that new entrants (especially women) receive harsh beatings and have their wares destroyed. This discourages all but the most persistent. Once in, though, she said she was protected. The groups controlling the space do battle in defence of their 'members', even against small forces of City Council guards. Mary expressed a sense that they were unified against outside forces, despite feeling that one ethnicity dominated the others at Nyayo. The informal responsibilities to fellow traders also extend to mutual assistance when traders face fines in court. Another seller, Kimani, said that the Nyayo hawkers work hard to protect their good corporate name. They lash out at any threats: outsiders who try to take advantage of the jam to steal from cars or from sellers, as well as sellers who deal improperly with customers. With pride, he told me about having beaten a thief almost to the point of death. In this way, the hawkers assert a claim over the space, making it a kind of territory. Their claim, however, exists in an uneasy relationship with the more formal regulation of Nyayo by City Council.

That traffic, like the roadside, can become a worksite has something to do with the fact that road space is understood to be available for unsalaried labour. The Harvard Project On the City writes that in Lagos the increased privatisation of space in the city means that roads and roadsides (and other infrastructural undersides) get claimed for a range of activities because they maintain a residual publicness. This sense of publicness also means that regulation regimes governing road space are ambiguous. As a child once said to me, 'the road doesn't belong to anyone, it belongs to God'. In Ghana, Gabriel

Klaeger describes bread sellers who find a similar freedom on certain kinds of roads: these sellers choose to work on suburban roads rather than city centre streets because these are a kind of no-man's land, not subject to the strict regulation of other spaces (Klaeger 2012: 542). The fact that Nyayo houses a hundred hawkers suggests that the space is more available for trading activity than the Central Business District, where hawkers are chased away aggressively.

Nevertheless, hawkers' sense of a right to work at Nyayo does not mean that they are left in peace. If the road space here is a no-man's land, this also means that the hawkers' territorial claim also cannot be fully established. While drivers' use of the road is regulated by traffic police, hawkers' use of roads and roadsides is managed by city council workers. The city's government, now called Nairobi City County (NCC), is the main body regulating the use of public space in Nairobi. The 'fertile' ground of roads and roadsides is a prime source of income for this body. Whereas traffic police skim an excess off circulation by blocking it in roadblocks, NCC is embedded in circulation itself. Indeed, one could almost say that *askaris* are a part of the city's infrastructure. These officials operate through a form of taxation: the daily accumulation of tiny pieces of profit gleaned from the already tiny profit margins of unwaged work. Most occupants of roadside space pay regular fees to NCC *askaris*, whether in the form of daily rent, bribes or fines. But mobile hawkers, because they are hard to pin down and because they work on the road itself, exist in a kind of grey area. Their territory is not defined enough for license fees to be charged, making it difficult to formalise their kind of business activity. As a result, *askaris* can only express their jurisdiction over the space through repression: by extracting bribes or arresting hawkers. Uniformed and plainclothes guards regularly raid Nyayo roundabout and pack hawkers into the backs of vans that bring them directly to court. There the hawkers are subject to large fines levied for contravention of city bylaws ranging from 'unlicensed trading' to 'dumping', 'general nuisance' and 'obstructing traffic'.

A game of cat and mouse therefore persists between city authorities and their forces and the sellers who assert their own claim to the roads and streets of Nairobi. Hawkers tacitly accept the state's legal claim on the public space; they seem resigned to these authorities' power to arrest and impose fines (or even jail time).[16] Even while understanding that such arrests are explicitly extractive, hawkers rarely get into overt confrontations with the state. Rather,

16 All the hawkers I interviewed spoke of the inevitability of arrest and arraignment in court. They also said there was no point in denying the charges read against them and described routinely pleading guilty and paying the fines – about $12 (KES 1,000) per charge for each of four charges, unlicensed trade, dumping, general nuisance and impeding traffic flow.

in returning time and time again, they treat arrest itself as a form of taxation, albeit violent and coercive, that wins them the right to exploit the road space. They did not perceive their activity as criminal, lamenting instead that impelled by destitution and invited by the marketplace that is the jammed roundabout, they had a right to extract what they could from the opportunity presented. And through their daily presence, and their numbers, they simply overwhelm the state's ability (and perhaps even the state's desire) to remove them.[17]

Conclusion

Traffic is not new in Nairobi. Almost as soon as it came into being, the city was full of cars. As Terry Hirst writes in *The Struggle for Nairobi,* by 1928 Nairobi had 5,000 cars, making it the most 'motor-ridden' town in the world at that time (1994: 65). Motorised transport facilitated segregation between 'Europeans' and 'Africans' by allowing Europeans to live in cooler, better-watered settlements far from the city centre. Today, as the city continues to sprawl, workers are increasingly dependent on motorised transport, the private car is still central to self-actualisation in the Nairobi imaginary and building new roads is still the preferred means of addressing transport issues. Despite city government's renewed attention to the problem of traffic jams (most recently, with the 'removal' of certain key roundabouts, Nyayo Stadium among them), the jam will continue to loom large in the lives and imaginations of Nairobians.

Traffic of course represents more than just the number of cars in the city or even spatial settlement patterns. It also makes visible the rapid population increase in Kenya's largest city. In the post-independence period, urban migrants expected to find formal work in the city, and many did. The matatu industry arose to ferry these commuters between work and home spaces that were,

---

The alternative was two weeks in remand jail and a trial court unsympathetic to hawkers, with a very high probability of paying the very same fines, or higher.

17 The Kenyan state seems split on what to do with the persistent hawkers. Elected members of parliament (particularly around election time) tend to side with hawkers, making arguments about the lack of employment alternatives and the honesty of hawkers' labour, against executive authorities that prefer formal tax and levy paying businesses. Most Kenyan cities have large, formal, state-constructed spaces for hawker resettlement. But these tend to be taken up by wealthier traders – like the wholesalers who supply many of the Nyayo hawkers – and not by their intended beneficiaries.

according to persistent colonial logics, separate from one another. But many did not find the employment they hoped for, giving rise to the 'self-help city' (Hake 1977). Even as the hope of formal employment has faded, the city continues to entice millions from across East Africa. This has resulted in a productive and versatile unwaged economy. And many of Nairobi's new residents find themselves, at one time or another, working on the road.

City authorities, nervous about traffic's effect on the city's reputation and its 'productivity', often seem to envision the jam as the encroachment of messy informality onto road space which is properly clear. Attempts to clear this space through fines and violent displacement (and, sometimes, provision of alternative spaces) dovetail with other solutions for traffic jams; visions of the future which also scrub out the informality that characterises the lives and livelihoods of roadside vendors and matatu crews. In the search for 'efficiency' and a good image for investors, something is surely overlooked – namely, that the mass of urban entrepreneurs earning a living from the road (and especially from interruptions of mobility) *are* the city. The forms of measurement within which the jam is understood as a space of waste simply do not take this population into account. To speak of wasted man-hours in a city in which seventy per cent of the population is not formally employed has the quality of the absurd. When most Kenyans refer to 'idleness', it is not 'losing time' that troubles them but rather the fear of no work at all – empty time.

Nevertheless, this chapter has tried to take seriously the kind of time that is evoked when authorities speak of 'wasted man-hours' in the jam – time defined by the structures of salaried labour – and to demonstrate the ways in which it is also destabilised in the jam. Looking at time as it actually is in the city, rather than as it is supposed to be, suggests that living or working in Nairobi requires, even generates, a certain flexibility with respect to time. Time on the road is not linear; even time-space compression, classically an effect of motorised transport and the development of capitalism, can become time-space *extension* when one spends an hour moving five hundred meters. For salaried workers, the jam makes time less regular, less controllable. Conversely, while the irregular time-patterns of 'informal labour' have been a means of distinguishing it from 'formal' labour, those working in the jam find their time structured by the daily and monthly tempos of salaried labour. Indeed, in the jam, salaried and unsalaried workers find themselves constrained, together, by a form of time that belies the association of motorised modernity with speed and efficiency. Nairobi residents imagine that time should be measured and structured, an imaginary shaped by industrial capitalism and an ethic of hard work. But the life of the jam suggests that this moral imaginary might be out of step with the reality of contemporary urban life.

References

African Development Bank. 2010. *Tracking Africa's Progress in Figures*. Tunis: African Development Bank Group.

Bélanger, Pierre, Chuihua Judy Chung, Joshua Comaroff, Michael Cosmas, Sonal Gandhi, A. David Hamilton, Lan-Ying Ip, Jeannie Kim, Gullivar Shepard, Reshma Singh, Nathaniel Slayton, James Stone and Sameh Wahba. 2001. 'Lagos: Harvard Project on the City.' In *Mutations: Événement culturel sur la ville contemporaine*, edited by Rem Koolhaas, Sanford Kwinter, Stefano Boeri, Nadia Tazi and Hans U. Obrist, 650–719. Bordeaux: ACTAR.

Coetzee, J.M. 1988. 'Idleness in South Africa.' In *White Writing: On the Culture of Letters in South Africa*, 12–35. New Haven: Yale University Press.

Conley, Jim. 2012. 'Sociology of Traffic: Driving, Cycling, Walking.' In *Technologies of Mobility in the Americas*, edited by Phillip Vannini et al. (eds), 219–236. Oxford: Peter Lang.

Cortázar, Julio. 1973. 'The Southern Thruway.' In *All Fires the Fire and Other Stories*, 3–29. New York: Pantheon Books.

De Boeck, Filip. 2012. 'Infrastructure: Commentary from Filip De Boeck.' *Cultural Anthropology Online*. November 26, 2012. www.culanth.org/curated_collections/11-infrastructure/discussions/7-infrastructure-commentary-from-filip-de-boeck.

Evans-Pritchard, Edward E. 1939. 'Nuer Time-Reckoning.' *Africa* 12 (2): 189–216.

Gidwani, Vinay. 2008. *Capital, Interrupted: Agrarian Development and the Politics of Work in India*. Minneapolis: University of Minnesota Press.

Goffman, Erving. 1971. *Relations in Public: Microstudies of the Public Order*. New York: Basic Books.

Guyer, Jane. 2004. *Marginal Gains: Monetary Transactions in Atlantic Africa*. Chicago: University of Chicago Press.

Hake, Andrew. 1977. *African Metropolis: Nairobi's Self-Help City*. New York: St. Martin's Press.

Harvey, David. 1989. *The Condition of Postmodernity: An Enquiry into the Origins of Cultural Change*. Oxford: Blackwell.

Hirst, Terry. 1994. *The Struggle for Nairobi*. Nairobi: Mazingira Institute.

IBM. 2010. *IBM Global Commuter Pain Study Reveals Traffic Crisis in Key International Cities*. Accessed July 18, 2016. www-03.ibm.com/press/us/en/pressrelease/32017.wss?.

Klaeger, Gabriel. 2012. 'Rush and Relax: The Rhythms and Speeds of Touting Perishable Products on a Ghanaian Roadside.' *Mobilities* 7 (4): 547–554.

Klopp, Jacqueline M. 2011. 'Towards a Political Economy of Transportation Policy and Practice in Nairobi.' *Urban Forum* 23 (1): 1–21.

Lamont, Mark. 2013. 'Speed Governors: Road Safety and Infrastructural Overload.' *Africa* 83 (3): 367–384.

Lefebvre, Henri. 2002. *Everyday Life in the Modern World*. London: Continuum.
Macharia, Keguro. 2010. 'Traffic in the City.' Blog post. Accessed July 18, 2016. https://gukira.wordpress.com/2010/11/12/traffic-jam-in-the-city/.
Manji, Ambreena. 2014. 'Bulldozers, Homes and Highways: Nairobi and the Right to the City.' *Review of African Political Economy* 42 (144): 206–224.
May, Jon and Nigel Thrift. 2001. 'Introduction.' In *TimeSpace: Geographies of Temporality*, 1–46. London: Routledge.
Mbugua, Ng'ang'a. 2011. 'Why Traffic Jams are Not a Laughing Matter.' *Business Daily* (April 26). Accessed July 18, 2016. http://allafrica.com/stories/201104270011.html.
McGregor, Sarah and David Malingha Doya. 2014. 'Traffic Costs Nairobi $570,000 a Day as No. 2 Africa Hub Clogs.' *Bloomberg*, April 12.
Mrázek, Rudolf. 2002. *Engineers of Happyland: Technology and Nationalism in a Colony*. Princeton: Princeton University Press.
Mrázek, Rudolf. 2004. 'Bypasses and Flyovers: Approaching the Metropolitan History of Indonesia.' *Social History* 29 (4): 425–443.
Mungai, Mbugua Wa, and David A. Samper. 2006. 'No Mercy, No Remorse? Personal Experience Narratives about Public Passenger Transportation in Nairobi, Kenya.' *Africa Today* 52 (3): 50–81.
Mutongi, Kenda. 2006. 'Thugs or Entrepreneurs? Perceptions of Matatu Operators in Nairobi, 1970 to the Present.' *Africa* 76 (4): 549–568.
Mwangi, Fred. 2014. 'Kidero Forms Team to Decongest City Traffic.' *The Star*, January 28.
Ndemo, Bitange. 2013. 'To Democracy We Must Add Discipline.' *Daily Nation*, December 16.
Ngigi, George. 2015. 'CBK Backs Cheap Oil Imports to Shore Up Shilling.' *Business Daily*, February 22.
Nixon, Rob. 2011. *Slow Violence and the Environmentalism of the Poor*. Cambridge: Harvard University Press.
Ross, Kristin. 1996. *Fast Cars, Clean Bodies: Decolonization and the Reordering of French Culture*. Cambridge: MIT Press.
Schivelbusch, Wolfgang. 1987. *The Railway Journey: The Industrialization and Perception of Time and Space*. Berkeley: University of California Press.
Shaw, Rosalind. 2002. *Memories of the Slave Trade: Ritual and the Historical Imagination in Sierra Leone*. Chicago: University of Chicago Press.
Slaughter, Joseph. 2004. 'Master Plans: Designing (National) Allegories of Urban Space and Metropolitan Subjects for Postcolonial Kenya.' *Research in African Literatures* 35 (1): 30–51.
Thompson, Edward P. 1967. 'Time, Work-Discipline, and Industrial Capitalism.' *Past and Present* 38: 56–97.

CHAPTER 4

Stories of the Road

Perceptions of Power, Progress and Perils on the Accra-Kumasi Road, Ghana

Gabriel Klaeger[1]

In the town of Kyebi, where I conducted research on the Accra-Kumasi road (AKR), residents told me a tale about Potroase, a nearby village jokingly referred to as 'Airport':

> A man once travelled in his car from Accra to Kumasi in the night. Upon reaching Potroase, the man realised that he was running out of fuel and decided to buy a few gallons. That night, the small village appeared to the traveller like a big city – or like an airport – with plenty of animation, modern lighting as well as a filling station. After paying for his fuel, the man continued his journey only to realise later that he had forgotten to get his change. He decided to collect it on his way back. But when he returned to Potroase, he was surprised to find neither lighting nor any filling station. For some mysterious reason, the place had turned back into what it usually is: a dull and underdeveloped village somewhere in the middle of a major trunk road.

Kyebi's residents narrated this tale for different reasons. Some simply wanted to explain why the small village located both sides of the main road had come to be nicknamed 'Airport'. Others were eager to draw my attention to the powers, such as spirits (*sunsum*) and witchcraft (*bayie*) that some of the residents (not necessarily the narrators themselves) assumed were at play in this village. It was possibly these powers that had transformed the village into a modern and lively place that night or that had tricked the traveller's perception and bereft him of his money. The malicious powers alluded to in the tale further served to interpret the numerous fatal motor accidents that had occurred on the Potroase road – a section regularly referred to as 'death trap' or

1 I am grateful to Kurt Beck, Julie Soleil Archambault, Richard Fardon, Sophie Mew and Michael Stasik for their comments on earlier versions of this work. Some of the issues presented here were discussed at the African History Seminar (Department of History, SOAS) in 2008. Doctoral research on which this chapter is based was supported by a Research Student Fellowship (SOAS) as well as by grants from the Central Research Fund (University of London).

 | DOI 10.1163/9789004339040_005

'death zone'. People also recounted the tale and referred to the road's dangers to explain why a new road section – a bypass – had recently been built. They assumed that the main purpose of this bypass was so that travellers on the AKR would no longer need to pass through the Potroase 'death trap'. Yet others claimed that the road's powers and dangers were merely a pretext for building the bypass – a pretext used by those who had promoted the bypass in order to weaken the political and economic position of Kyebi.

The bypass, called Apedwa-Bunso road or simply referred to as 'new road' (Map 4.1), was completed a few months before my arrival in Ghana, in early 2006. Most of the AKR's traffic was now diverted away from not only Potroase, but also Kyebi, the capital of the Akyem Abuawka kingdom. Among my friends and other Kyebi residents, the 'removed' road triggered heated debates about the diverse effects this road had on their community. At the centre of the conversations that I followed closely were a series of stories, often in the form of rumours. Similar to the 'Airport' tale, they geared towards who and what had promoted, or failed to prevent the building of the new road. My aim in this chapter is to examine the stories, beliefs and speculations

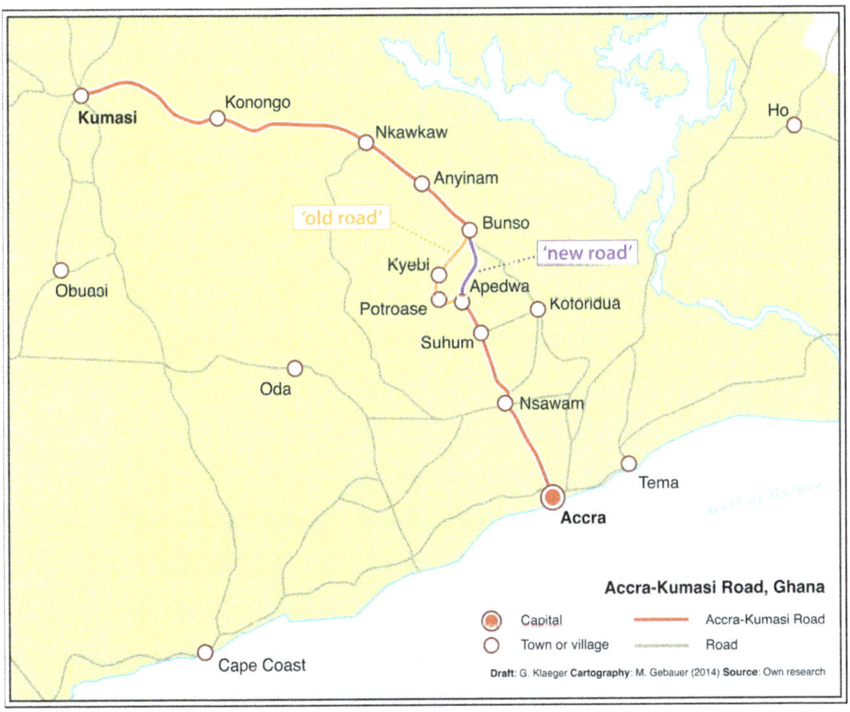

MAP 4.1 *The 'new road' (Kyebi bypass) on the Accra-Kumasi road*

that were brought forward by my informants at the decisive historical moment in the life of this community. I argue that these stories and rumours are social commentaries, namely narrative devices for articulating people's concerns about the town's recent detachment from the main AKR. Moreover, these narratives evoke particular imageries that 'make' the AKR – namely as a route for and realm of power, progress and perils. Thus, the stories and imageries provide deep insights into the ways in which Kyebi residents perceive and experience the AKR as embedded in the wider contexts of politics, development and uncertainty. These perceptions resurface among people who are tangibly concerned with roads and road-building in various parts of the world (see, for instance, Dalakoglou 2010; Harvey 2005; Lamont 2013; Masquelier 2002), and whose road experiences are framed by a sense of ambivalence and instability.

Asante Great-roads versus Akyem Authority

Social scientists have described roads as means and contested objects in the struggle for political control and power (Alber 2002, Campbell 2012, Fairhead 1993, Havik 2009, Ispahani 1989, Wilson 2004). The concept of roads as 'paths of authority' (Fairhead 1993) was also reiterated among certain people in Kyebi who associated the construction of the new road with a power struggle along ethnic lines. It was rumoured, for example, that the Asantehene – the king of the Asantes with his seat of power in Kumasi – had urged the country's president at that time, John Agyekum Kufuor, to build the bypass. Of course, Kufuor, an Asante himself and loyal to the Kumasi elite, had complied. But why was the Asantehene so keen on having that road built? Those who believed in the conspiracy theory explained that it was all connected to familiar themes regarding the long-established 'tribal' rivalry, even enmity, between the Asantehene and the Okyenhene, the king of the Akyems. The rivalry goes back to past inter-ethnic wars during one of which the Asantehene Osei Tutu was killed by the Akyems in 1717 (Fynn 1973: 64). Allegedly, since then, successive Asante kings have always sought to avoid passing through Akyem territory, or at least through Kyebi, the Akyem capital. The construction of the new road now enables the current Asantehene not only to travel more quickly to Accra, but also to bypass and thereby ignore Kyebi. It could therefore be argued that the realigned AKR has an Asante-dominated route again. For Kyebi residents, especially those close to the royal palace, the new road stands for an attempt to symbolically weaken the Okyenhene's authority over 'his' road sections and to contest Akyem's political position more generally.

Asante Power and the Pre-colonial Great-roads System

My informants who stressed the current ethno-political role of the AKR were generally aware, but had little detailed knowledge of how the road has been historically embedded in the often conflictual relationship between Asante and Akyem. A comprehensive account of the road's connection to historical processes of power and conquest is provided by Ivor Wilks (1975, 1992). Drawing mainly from colonial and missionary sources, Wilks describes how, at the beginning of the 19th century, the Asante capital of Kumasi had 'free and unobstructed communication with all the leading provinces, by roads or paths, which [...] are collectively called great roads' (Dupuis 1824: xxvii; cited in Wilks 1975: 1).[2] In Twi, the 'great roads' were called *akwan-tempon*, from *ekwan-pon* (great-road) and *ten* (straight). Wilks identifies eight great-roads (Map 4.2) that include four northbound 'inland roads' (Route V–VIII) and four southbound 'great maritime causeways' (Route I–IV).[3] Route V is of particular interest here. It corresponds to the so-called 'Akyem route' (Yarak 1990: 123) that led through the Akyem kingdom to Accra and that strongly affected the course of the Accra-Kumasi motor road built at the beginning of the 20th century (see below).

The great-roads system formed a network with a radial design, organised according to one central node, namely the Asante capital of Kumasi. The system was designed to achieve two different ends: 'to promote the flow of trade, and to facilitate the maintenance of political control' (Wilks 1975: 16).[4] Political control is the dominant context in which the great-roads system is placed. Wilks (1975: 18) stresses the fact that the evolution of the roads system was closely related to the process of political incorporation of conquered territories in Greater Asante, yet that this process was not always a smooth one. He argues that, 'opposition to the central government's road building programme was one of the principal features of the syndrome of resistance to its imperial

2 Wilks' sources were, among others, the British Consul Dupuis (1824), Bowdich (1819) and the Basel missionary Christaller (1881). The geographer Dickson (1961: 33–34) who describes the early southern routes in his treatment of the development of road transport mainly draws from Dupuis' account.

3 It is further noted that the great-roads system was maintained by the central government in distinction from the smaller and local road networks that were under the responsibility of district authorities (Wilks 1975: 1).

4 The Asante great-roads are depicted as 'instruments for the maximization of economic benefits (trade and production)' and as 'instruments for the maximization of political control (government)' (Wilks 175: 26); see also Yarak (1990: 115–126). Wilks (1975: 13–14) complains that Gould's (1960) geography of transportation in Ghana shows the development of the communication network in the 20th century, but fails to consider its 19th century precedent.

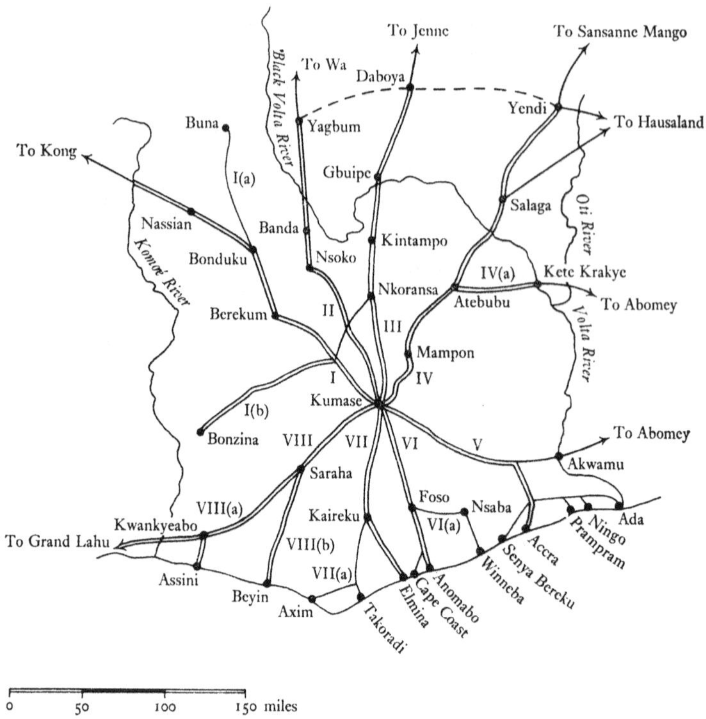

MAP 4.2 *The Asante great-roads network in the early 19th century*
SOURCE: WILKS 1975: 11

expansion, just as the closure of existing roads became one of the earliest indications of rebellion' (Wilks 1975: 25).[5]

After the eight great-roads had been established in the early 19th century, apparently no further major highways were built. Indeed, some of them fell into disuse with the growth of British interest in the southern provinces of Asante and with the establishment of the Gold Coast Colony and later the Asante protectorate. This meant that Kumasi lost control over sections of the four southern great-roads. To hinder communication between Asante and Gold Coast towns, the British administration even closed some of the routes in times of conflict. The overall great-roads system ceased to exist when Greater Asante was occupied by the French, Germans and British at the end of the 19th century (Wilks 1975: 12–13).

5 Wilks (1992: 175) stresses again that creating the Asante road system involved the negotiation of numerous agreements with local chiefs through whose lands the roads were routed; the establishment of chains of appointed halting-places; and the setting up of control posts manned by the 'highway-police' (*nkwansrafoɔ*) at strategic points.

Akyem's Claim to Authority over the Road

One of the instruments of pre-colonial Asante authority and rule was the aforementioned 'Akyem route' (Route V). It was at the core of the incorporative process of Akyem into Asante, and it was central to the recurring power struggles between the two kingdoms. For instance, in 1742, following the Asante invasion and the capture of the Akyem royal households, an agreement to guarantee Asante protection to the Akyem countries and exemption from tribute was on the verge of being negotiated. In return, free passage through the Akyem countries to Accra was to be granted to the Asante. But Akyem Abuakwa refused, and the attempt to open a direct road – the later Route V – through Akyem territory to Accra had to be abandoned at this time (Wilks 1975: 25). It was only after the incorporation of Akyem Abuakwa into Greater Asante in 1766 that the route was finally opened (Wilks 1975: 28).[6] The road ceased to function as an instrument of Asante domination with the establishment of British colonial government in Accra, the centre of power thus emerging at the other end of the road. After 1957, the Government of Ghana became the new (official) authorities of the former Asante trade route. With their ministries, agencies and political and developmental agendas, they gradually transformed the route into today's existing motor road.

However, the rumour about the current Asantehene's influence on the bypass construction shows that people in Kyebi continue to regard the road as charged with ethnic ('tribal') and political authority. With respect to the section of the AKR that crosses Akyem Abuakwa territory, they claim that 'it belongs to the Okyenhene' and that it is under the political and symbolic authority of the traditional ruler. To provide supporting evidence, some refer to the historical fact that during the colonial era, one important section of the road (the Apedwa-Kyebi section) was transformed into a motorable state by the then king, Nana Sir Ofori Atta I (see below). A more contemporary piece of evidence, in people's view, is a huge billboard (Fig. 4.1). It was erected some years ago near the AKR road in Nsawam-Adoagyiri, where the road crosses the Densu river that marks the boundary between the Akyem and Akuapim kingdom. Its bold letters read 'Welcome to OKYEMAN – The Kingdom of Akyem Abuakwa'. On the left is a close-up picture of a roaring leopard, which stands for the king of all the animals formerly found in the dense Atiwa forests of Akyem (*Kwaebiribirim*), which travellers pass through on their way to Kumasi.[7]

6 See Affrifah (2000: 78–80; 102–103) for the crucial role of trade routes in the pre-colonial Asante-Akyem relations.

7 The leopard as a symbol of power also figures on the widely known state emblem of *Okyeman* (literally: the state of Akyem).

FIGURE 4.1 *Entering the Akyem kingdom in Nsawam-Adoagyiri*
SOURCE: G. KLAEGER, 2007

People add that the Okyenhene's authority over the road is regularly manifested through ritual means. When travelling from his Accra residence to Kyebi for festive occasions, for example, the Okyenhene's car stops at Nsawam for one of his spokesmen (*akyeame*) to pour a libation by the roadside. I was told that praying to the deity of the river at the entrance of the kingdom is a way of showing the deity her due respect, but also reconfirms the King's sovereignty over the land that he is about to enter via the road. As I have shown elsewhere (Klaeger 2013a), such ritual road performances occasionally allow turning the road into a public space for displaying chiefly power and authority.

These symbolic markers and practices nurture people's view of the road as a path of authority and a realm of power. The idea that the Asantehene may have been able to maintain control over a road section in Akyem – and contest the sovereignty of the Okyenhene, his rival – makes a similar point. Some of my interlocutors reasoned further that the Asantehene's move must have been motivated by his desire to weaken the Akyem capital both politically and economically. It is to this and to closely related suspicions held among frustrated Kyebi residents that I now turn.

(Dis)connecting Pathways to Progress

'They want to destroy the town', I often heard people say in Kyebi. They claimed that the 'new road' was meant to make the town and its residents 'suffer', since it disconnected them from the main road and, most importantly, from the rewards of passing traffic. But whose intention could this have been? While some held the Asantehene responsible (see above), others looked to Accra and blamed the central government that is believed to continue marginalising the Akyem capital politically and economically, as it has done in the past.

The British colonial government, for instance, is remembered to have abandoned the royal capital following the sensational 'Kibi murder case'.[8] The case arose in 1944 after the sudden disappearance of one of the sub-chiefs to the then Okyenhene shortly after his death. The sub-chief had allegedly been murdered in Kyebi during a sacrificial ritual for the King's burial.[9] The murder case generated a lot of attention in the Gold Coast colony and beyond and led to the stigmatisation of the town as backward, uncivilised and even dangerous. Such stigmatisation is recalled to this day by the popular phrase that 'in Kyebi, they cut heads' (Klaeger 2007a: 112). As a consequence of the politicised (and in fact unresolved) murder case, the colonial government excluded the town from some of its investment and development initiatives. The situation did not improve in the years after Ghana's independence, when one of the town's most prominent figures and a member of the royal family, the scholar and politician J.B. Danquah, became the main adversary to Kwame Nkrumah, Ghana's first president (Rathbone 2000). The fact that Kyebi was Danquah's hometown and a stronghold of anti-Nkrumahism and thus anti-government sentiments did not redound to the town's advantage.

I was nonetheless surprised by people's claim that the government in power headed by President John Kufuor was taking an anti-Kyebi stance. After all, Kufuor's party, the ruling New Patriotic Party (NPP) is strongly supported in Kyebi. This is linked to party-historical reasons and to the fact that Nana Addo Dankwa Akufo-Addo, at that time one of the leading NPP members and the Minister of Foreign Affairs, is a native of Kyebi. Yet it was precisely the relationship

8 Kibi (occasionally also: Kibbi) was the British notation for Kyebi; both versions are used today in written and spoken language.

9 See Rathbone (1989, 1993) for an historical account and the political circumstances of the alleged ritual murder that occurred in Kyebi in February 1944, after the death of the Okyenhene Nana Sir Ofori Atta I, and that provoked political and media attention beyond the Gold Coast colony.

between Kufuor and Akufo-Addo that was questioned and became the subject of a road-related rumour. It is a well-known fact that both politicians had aspired to the post of the NPP's presidential candidate in 2000 and 2004 and that Kufuor won the race both times, within the party as well as for the presidential elections. It was suggested that Kufuor had arranged for the new road to be built so as to reaffirm his superiority over Akufo-Addo. His internal rival would be demoralised and humiliated by 'removing' the main road, thus weakening the economic activities in his hometown.[10] What these allegations hint at is not just political rivalry, but also socio-economic decline as one of the (intentional) consequences of infrastructural disconnection. The following sections bring to light how the road is perceived as a pathway to progress and development and how its absence or abandonment can be disadvantageous to the community affected, a perspective that is explored by several scholars (Harvey 2012; Harvey and Knox 2012; Colombijn 2002).

The Kyebi Road: A History of Neglect and Frustrations

In Kyebi, aspirations for motor road access on economic grounds seem to be nothing new, and neither are the frustrations that town residents currently experience in this respect. This is confirmed by several historical sources such as those retrieved from the Akyem Abuakwa State Archives.[11] In a letter dated 1921, a colonial administrator residing in Kyebi reported that the town residents 'have had many disheartening experiences with this road'.[12] He referred to the fact that despite the so-called 'road revolution' triggered by the growing availability of motor vehicles in the early 20th century, Kyebi was struggling to obtain, and to maintain, viable motor road access.

The former Akyem trade route was already reported to be in an unsatisfactory state around 1850. The geographer Kwamena Dickson (1961) argued that one factor was the rapid growth of the vegetation, another was the apparent lack of interest of the British and Dutch administration in road construction. In 1870, the subject of roads was debated in the Legislative Council, concluding

10 Various commentators claim that the NPP-internal rivalry between Kufuor, an Asante, and Akufo-Addo, an Akyem, is influenced once again by 'tribalism'. The contest between the two came to an end when Akufo-Addo finally became the NPP flagbearer in 2008, following the withdrawal of President Kufuor who was not allowed to serve a third term in office. The NPP lost the 2008 elections against the National Democratic Congress (NDC) led by John Atta Mills. See Nugent (2001a, 2001b) on the 2000 elections and Jockers et al. (2010) on the 2008 elections.

11 The Akyem Abuakwa State Archives (AASA) are based at the *Ofori Panyin Fie*, the Okyenhene's palace and administrative seat in Kyebi.

12 AASA.10.45: Ass.DC to DC (3 November 1921).

that 'good carriage roads were too expensive to build and were, in any case, undesirable' (Dickson 1961: 35); it was argued that 'there was no prospect of concentrated traffic that would repay the public for making good roads' (Dickson 1961: 35) and that improving the existing pathways would be sufficient, the natives favouring head-loading anyway.[13] A further objection was that 'the Administration had no power to carry out road building schemes' and finally, that '[t]he natives themselves did not want good roads, because an enemy might use them in time of war' (Dickson 1961: 36).

According to Dickson, the turning point was the introduction of motor vehicles at the beginning of the 20th century. In 1901 the Governor of the Gold Coast colony pleaded for 'the building of roads good enough for motor cars and traction engines. Such roads [...] would be easy to keep clear, and the use of motor vehicles would free, for other occupations, a large proportion of the country's labour force' (Dickson 1961: 36). Several roads were therefore constructed or repaired for the use of motor traffic – one among them was the road leading from Accra through Nsawam and Apedwa to Kyebi. In 1905 the *Goldfields of Eastern Akim Ltd*, with government assistance, completed the six yard wide motor road built specifically to reach areas of expanding cocoa production and gold mining in the Akyem area (Dickson 1961: 37; Gould 1960: 15, 38). The laterite-surfaced road was, however, quickly ruined as it could not stand up to heavy motor cars and lorries.[14]

By 1915, the colonial government published plans to reconstruct the drastically dilapidating Nsawam-Kyebi section, but to the great disappointment of local residents, the road works ended before they even reached the Akyem capital. In a letter sent from his palace, the Okyenhene Ofori Atta I complained:

> It was extremely unfortunate that the Government found reasons to alter the original plan which resulted in the discontinuance of the reconstruction from Apedwa to Kibbi [despite] the growing importance of the road which is evidenced by a number of villages now in course of construction.[15]

The 'discontinuance' of road works by the British during (and after) World War I was a common phenomenon. It was connected to the war crises, lack of funds

13 A quite similar, and rather odd, position is taken by Moriarty and Beed (1989) with regards to present-day transport in 'tropical Africa'.

14 The 'complete revolution in road transport in the country' was the arrival of the light American Ford lorries in 1913 (Heap 1990: 38–39); also see Gewald (2002), Gewald et al. (2009), Gould (1960: 66) and van der Laan (1981: 555).

15 AASA.10.20: Omanhene to DC (16 April 1917).

and to British officials' relative neglect of roads in the Gold Coast in favour of railways (Wrangham 2004: 14).[16]

For several years, the road section leading to Kyebi remained neglected by the government and hence became 'quite impassable for Motors'.[17] By 1916, Ofori Atta I became exasperated. The King decided to make the Apedwa-Kyebi road fit for motor service at his own expense, as this was 'one of his paramount interests' and was expected to be 'a great acquisition to his people and himself'.[18] His project triggered great excitement, for instance, Mr. Müller from the *Basel Mission Factory* in Nsawam heartily congratulated the King and predicted that 'Kyebi would in the near future be an important centre of business'.[19]

The widely anticipated boost to the local economy was, however, retarded once again. This time it was not the government, but the Italian contractor Mr. Dalberto (hired for the road works in 1916) who disappointed the King. Six months after signing the contract, Ofori Atta I himself inspected the Apedwa-Kyebi road section in his Studebaker car – and was not at all pleased. He wrote to Mr. Dalberto:

> I awfully regret to have to state with regard to the condition of the road as I saw it, that never have I seen a work so half done, or so haphazardly brought into the so-called completion as this road work. [...] [C]ertain parts of the road are so soft that every motor lorry is bound to sink seriously, thus causing considerable annoyance and trouble to the passengers.[20]

What angered the King was not only that 'it [would] be a matter of further heavy expense to make this road what it should be', but more particularly that 'owing to the bad condition of the road Firms [were] unwilling to send their lorries to Kibbi'.[21]

The struggle to incorporate Kyebi into the network of motor roads continued. In the years that followed, the King occasionally employed his own

16 According to Wrangham (2004), the colonial administration promoted a 'railway imperialism' that led to the complete neglect of the potentials of roads, actual transport needs and the growing use of motor lorries. Alber (2000: 288) observes a form of 'imperialism' with regard to road construction in the French colony of Dahomey (see also Boni 1999: 62–63). On the competition between railway and motor transport, see Dickson (1961: 39), Gould (1960: 22–76) and Heap (1990: 25).

17 AASA.10.8: Swanzy Ltd to Omanhene (16 September 1916).

18 AASA.10.8: Danquah to contractor (10 November 1916).

19 AASA.10.8: Müller (Basel Mission Factory) to Omanhene (19 January 1916).

20 AASA.10.20: Omanhene to Dalberto (10 March 1917).

21 AASA.10.20: Omanhene to Dalberto (10 March 1917).

labourers and asked for a government grant, advocating that his people 'had an "ardent desire" to improve roads but lacked equipment' (Wrangham 2004: 12). In 1921, after another period of absolute neglect, the King ordered his subjects to take part in communal labour that he supervised himself.[22] Upon completion of the intense road works, one government official reported that it was due to the 'greatest enthusiasm' displayed by the King and his people that the work had 'turned out to be most excellent'.[23] Ofori Atta I also called the road 'excellent and first class' and finally considered it suitable for motor traffic – ready for firms and commercial vehicles to come to Kyebi for business.[24]

The partially 'African-constructed road system' of the Gold Coast (Wrangham 2004: 14), now also reaching Kyebi and neighbouring towns, enabled cars and lorries to enter in great numbers to the heart of Akyem for both commercial and administrative use. Rathbone (1993: 13) notes that with improved road transport 'the physical space of Akyem Abuakwa seemed to shrink as journeys, which had taken days on foot, now took hours by motor transport'. Thus the infrastructural development not only impacted strongly on local economies predominantly based on the cultivation of cocoa, but also 'helped reinforce Akyem Abuakwa's identity' (Rathbone 1993: 13) more generally.

In the decades that followed, Kyebi continued to be one of the stops along the AKR. The road partially deteriorated during the Great Depression, but it was newly surveyed and completely realigned by 1938 (Gould 1960: 72). Further major reconstruction exercises were undertaken in the late 1950s, 1970s and early 1990s, none of which affected Kyebi's access to the AKR. The most recent reconstruction of this road started in 2002 and was ongoing at the time of writing. It included the alteration of the AKR's course around Kyebi where the already mentioned bypass was built and completed in early 2006.

Road (Re)constructions and the Promise of Speed

After almost a century of direct access to one of the country's major traffic corridors, people in and around Kyebi feel that their established roadside communities have suddenly become isolated again. While rumours and allegations

---

22 Wrangham (2004: 1, 6) discusses local road building initiatives as an 'African response to a lack of colonial government provision', leading to what she calls an 'unexpected African road revolution [as] striking evidence that Africans were not merely passive subjects of colonial rule'. For similar initiatives in other places, see Boni (1999), Drummond-Thompson (1992) and McCaskie (2001: 128). Heap (1990) rather emphasises the joint initiatives of locals and colonial administrators.

23 AASA.10.45: Ass.DC to DC (3 November 1921)

24 AASA.10.45: Omanhene to DC (1921 [not dated])

suggest that the road was intentionally 'removed' to 'destroy' the town, the government's activities suggest a different, if not contrary agenda that requires elaboration.

The bypass was one element in the reconstruction and partial dualisation (upgrading to dual carriageway) of the entire 222 km long AKR.[25] These works in themselves formed a core element of a large-scale project, the Road Sector Development Programme (RSDP). The RSDP was administered by the Ministry of Roads and Transport (MRT) between 2002 and 2008. It was jointly funded by the Government of Ghana and a range of international development partners and donors, including the World Bank, which provided a credit worth US$ 220 million (World Bank 2008: 8).[26] In total approximately US$1.2 billion was allocated for the RSDP for maintenance, rehabilitation, reconstruction and road safety works on Ghana's entire road infrastructure (Ministry of Transportation 2007). The RSDP agenda was built on the understanding that transport is 'a catalyst for growth and development' (Ministry of Transportation 2007), thus the MRT was confident that their programme would '[e]nhance the operational efficiency of the road network to promote economic growth and the delivery of social services' (World Bank 2008: 4). In line with this, the RSDP was intended to support the Poverty Reduction Strategy set up by the Government, based on the assumption that poverty reduction directly related to accessibility to markets and services (Ministry of Roads and Transport 2003: 108).

The often-noted aim of road building (or at least upgrading) towards connectivity, progress and prosperity (Khan 2006; Árnason et al. 2007; Harvey and Knox 2012) was highlighted by stakeholders involved in the infrastructural project. For instance, the Ghanaian transport engineering company *COMPTRAN* described the road works as part of the Government's goal of 'poverty reduction, regional integration and economic development by improving accessibility of land-locked countries (Burkina Faso, Mali and Niger) to the ports of Ghana' (COMPTRAN n.d.). The company emphasised that the AKR works would allow for faster travel and therefore lower transport costs for passengers and freight, as well as bringing other economic and societal benefits. Roadbuilding was associated here with the 'promise of speed' that Harvey and Knox (2012: 523) describe as one of the 'enchantments of infrastructure'.

---

25 For this purpose, the road was divided into six sections: Accra/Ofankor-Nsawam (17.6 km), Nsawam-Apedwa (41.6 km), Apedwa-Bunso (23 km), Bunso-Anyinam (11 km), Anyinam-Konongo (85 km), Konongo-Kumasi (44 km; see Ministry of Roads and Transport 2004).

26 Other donors were several governments, development agencies, development banks and funds in Africa, Asia, Europe and the Middle East.

Speed and acceleration were recurring and at times critical themes in the process of the AKR reconstruction. For instance, when the modern Ofankor-Nsawam dual carriageway was inaugurated in June 2006, President Kufuor not only praised the Chinese contractors for the quality of the work carried out, but also for the fact that it was completed two months ahead of schedule (*Accra Daily Mail* 2006).[27] Nonetheless, works on most of the six road sections took much longer than anticipated so the reconstruction of the AKR was not completed when the RSDP ended in 2008. This result annoyed (and continues to do so) all those regularly driving and travelling on the AKR. Drivers and passengers are increasingly frustrated by dug-up roads, dust, mud, traffic jams and 'go-slows' and remain unconvinced of the nation-wide acceleration, progress and prosperity that the project claimed to promote.

Speed was also at stake during the construction of the new bypass, the Apedwa-Bunso road section. In mid-2002, when signing the work contract with an Israeli construction company, the Minister of Transport declared that the project was scheduled to be completed in 24 months, but 'urged the contractors to "fast track" it to 18 months' (*Ghanaweb* 2002). The company's representative pledged to complete it within twenty months depending on weather conditions (*Ghanaweb* 2002). In the end, the road was opened few months *behind* schedule in early 2005. What is more, the works were unsatisfactory as small cracks started appearing in the surface and very soon repair works became necessary.

The planners were able to implement their concern with speed more effectively with respect to reducing travel time on the AKR. The most significant contribution to this aim was the new bypass (22 km) now cutting ten km from the journey on the old Kyebi road (33 km). Both officials and road users now consider journeys to be faster because the AKR is shorter, but also because it avoids the steep hills and the dangerous bend around Potroase. Furthermore, traffic can move at higher speeds thanks to the new pothole-free road surface and the dual carriageway realised on some of the AKR's sections. Despite its often fatal consequences, acceleration has been promoted, even praised, as a vital component of progress, nation-building and modernity more generally. This nexus seemed key to the Ghanaian government's agenda to literally speed up economic development (see Khan 2006: 94).[28] It was also implied in the

27 The dual carriageway, including the flyover bridge at Ofankor, was built by the *Chinese Railway Engineering Co-operation*.

28 Several authors have observed how roads in other parts of the world have equally become entangled in this very nexus, for instance Árnason et al. (2007) and Harvey and Knox (2012).

FIGURE 4.2 *Promoting progress and acceleration for the nation*
SOURCE: G. KLAEGER, 2008

slogan used by the ruling NPP during its campaign for the presidential elections in 2008: 'WE ARE MOVING FORWARD'. The slogan alluded to the achievements of the Kufuor government, particularly with regards to road building and transport, as well as to the general drive and momentum that characterised – at least from an NPP perspective – the current state of the nation. Some of the huge NPP billboards urged to 'BELIEVE IN GHANA' while the statement 'WE ARE MOVING ▶▶ FORWARD' visually hinted at the 'accelerated nation' (see Árnason et al. 2007) by means of a red fast-forward symbol (Fig. 4.2).[29]

Moving Backwards and Slow Business

It appeared that residents in Kyebi were not all that enthusiastic about the government's dromocentric agenda.[30] This became apparent when I visited

29 A similar message was carried by the highlife song *Go Go High* (by gospel singer Philipa Baafi) that became the NPP electoral campaign song and a nation-wide hit.

30 'Dromocentric' is my attempt to denote both people's concern with and their practices involving speed in the sense of rapidity, a term which is derived from the Greek *dromos* (running, race, race course). *Dromos* is used in a similar vein by Paul Virilio (1986) who

the town in 2008, a few months before the presidential elections. Some of my friends scoffed at the optimistic slogan on the huge NPP billboard erected a few steps from the now quiet old road and joked about the irony. The only welcome aspect of the billboard was the image of Nana Akufo-Addo, pictured with a bright smile: finally, Kyebi's big man in politics had become the party's presidential candidate. Moving forward with Akufo-Addo as the new head of the government seemed a brilliant prospect. In town, many young men especially were confident that his victory would open up new job opportunities for his supporters and kin.

When I arrived in Kyebi in 2006 to carry out my fieldwork, people already had a strong sense that far from moving forward, the town was rather going backwards. Many of my acquaintances argued that business in town had gone down – 'by 70%', someone estimated boldly. People kept telling me that 'Kyebi is quiet, too quiet' and complained that 'our place has become dull, like a proper *kurase* (village)'. Others suggested that the town had 'collapsed', or become 'like a cemetery'. 'IS KYEBI DEAD?' a young man asked in a discussion thread on *Facebook*, claiming that 'Kyebi is now suffering economically' from the new paradigm of highway development that avoided towns as thoroughfares. 'Now the town is dying if not dead' he commented further. As a local politician warned in a public address, Kyebi was turning into a 'ghost town' (*Ghanaweb* 2005a).

Many residents claimed they were feeling the impact of the diverted traffic. 'The road, it meant a lot. The place was lively, so lively, business was booming', a chop bar owner called Auntie Grace told me. Although the great majority of travellers passing through Kyebi in the past did not stop, some still needed a break and looked for food, drinks and provisions, such as the cargo truck drivers from Kumasi, Burkina Faso or Mali. At times they needed help from local mechanics or vulcanisers and remained by the roadside for several days. Very occasionally, entire busloads of travellers descended upon Kyebi, such as when a Benz bus or STC coach broke down. These transients were valuable customers for those selling and working in stalls, chop bars, drinking spots, telecommunication centres and workshops, all located along the through road and particularly in the environs of the lorry station (see Klaeger 2012; Stasik 2013).

Before the bypass was built, Auntie Grace was able to sell two large saucepans of palm nut soup with meat to travellers per day. 'Now business is very slow' she said, and explained that she prepares only one smaller saucepan, mostly for local workers. The bakery behind the lorry station was obliged to

coined the terms 'dromology' (the science of speed) and 'dromocracy' (the condition of modernity with its every-increasing speed).

curb its daily production, while the mechanics and vulcanisers have little to do. Few of these small-scale entrepreneurs are as flexible as the fruit and vegetable sellers from neighbouring farming communities. Soon after the bypass was opened, many of them moved their stalls to the Apedwa Junction in order to gain access to the main AKR traffic, meaning more customers.

Frustration about the apparent decline in the economic activities of Kyebi was most tangibly expressed by a group of young shoemakers in their early twenties, who run an open workshop by the roadside opposite the lorry park. Appiah told me that 'definitely, we miss the old traffic, it used to make the town hot'. A few months earlier, they had taken white paint and written '*COME BACK STREET*' in bold letters on the road in front of their stall. By using the asphalt to display their frustration about the 'removed' road they wanted to express their hope that something could be done to make the road and roadside business more lively again – maybe with the help of the Okyenhene, their king, or Nana Akufo-Addo, Kyebi's big man in politics. Their request remained unanswered, just as their words on the asphalt slowly faded away.[31]

The fading words reveal how people view the road as potentially lively and economically active and as a pathway to progress and development. It was along similar lines that the Ghanaian government promoted their infrastructural project to link acceleration, movement and progress. For people in Kyebi, the reverse seemed to be taking place. This accounts for the rumour that all this could not have been accidental. Some suggested that denying Kyebi its laboriously acquired access to the main AKR was an intentional means employed by President Kufuor to make his adversary's hometown 'suffer'.

People have realised that there is more to the abandoned through road than merely its drawbacks referred to above. The change has brought some welcome improvements. Residents in the 'cut-off' roadside communities appreciate that reduced traffic has made living and working alongside the road much calmer and safer (see Klaeger 2013a). Some have told me – and this leads to yet another road rumour – that the prospect of an accident-free road stretch had actually pushed the Okyenhene to secretly give his support for the new bypass. Could it be true that the King had been in favour of the redirection of 'his' road, to the disadvantage of the royal capital? Some of my interlocutors did indeed assume such a connection. Rumours circulated that the King's concerns were specifically related to the accident-prone road section at Potroase, or 'Airport', the small village near Kyebi referred to at the beginning of this chapter.

31 According to one informant, the Okyenhene had tried to prevent the construction of the bypass at an early stage 'to avoid humiliation'. Similarly, Urry notes that 'the redirection of a path, or its elimination with a new road, will often be viewed as vandalism against [the affected] community' (2007: 32).

The Animate Road: The Causes of Recurring Crashes

'The Okyenhene wanted the accidents to vanish, once and for all' I was told. Some people in Kyebi claimed that the King was supportive of the views of road engineers and road safety experts, or that he had been bribed. The experts maintained that the high number of accidents on the AKR was, in part, connected to poor road design and to drivers' inability to cope with it – most obviously so in the case of Potroase. They argued that redirecting the Kyebi road was crucial to reducing further accidents and fatalities. People also claimed that the Okyenhene welcomed such a measure for another reason, that he hoped that ending the carnage on the Kyebi road would finally put an end to the common talk about the presence of malicious powers in Potroase, such as the spirits and witchcraft alluded to in the 'Airport' tale. This would eventually free Potroase and the road to the Akyem capital from being stigmatised as perilous, backward and mysterious. What draws my attention here is less people's speculations about the King's involvement in redirecting the Kyebi road, rather, how these stories invoke various perils that different commentators perceive as the possible causes of road accidents.

Black Spots, Death Traps

Alongside economic and developmental agendas, it was the pressing concerns for road safety that triggered the reconstruction works on the AKR. One major cause of accidents was the road's poor condition – as maintained by road professionals and the general public at the time of doing research, but also by early observers over a century ago. Dickson (1961: 34), for example, refers to an account of the Wesleyan missionary Thomas Birch Freeman who wrote in 1859 that in all his travels 'he [Freeman] had seen no pathways worse than those in Akim. The traveller, he warned, could not take his eyes off the ground before him without risking a fall and broken bones'. Similarly, in his report about a journey from Kyebi to Kumasi undertaken in 1881, the Basel missionary David Huppenbauer (1888: 14, 18) complained about bad and dirty paths. They were so narrow that travellers got caught in the brushwood and had their clothes torn to pieces. Due to obstacles such as fallen trees, water puddles and swamps, the paths could accommodate neither horses, donkeys nor carriages. Not even travelling in a hammock was an option as its carriers would stumble over roots 'as thick as a man' (*mannsdicke Wurzeln*) and drop the missionary. Huppenbauer writes that being carried was 'torture' and he was forced to walk throughout most of his sermon journey (*Predigtreise*; Huppenbauer 1888: 14). About a century later, Rathbone (1993: 7) describes the AKR as 'a much-ravaged road which degenerated into mud-slides in the rainy season and potholed dust-storms in the dry season', until it underwent extensive repairs in 1992. The road steadily gained a reputation as one of Ghana's worst and most dangerous roads and was labelled a 'death trap' by press commentators – a term equally used for other Ghanaian road sections with a high concentration of black spots and accident casualties.[32]

On the country's numerous death traps, one newspaper reported it being 'a serious national crisis because road traffic accidents are becoming very common and are robbing the nation of its valuable human resources' (*Ghanaian Chronicle* 2006b). In 2004, it was suggested that road accidents (estimated to claim around 1,800 fatalities every year) caused more deaths than AIDS in Ghana (*Ghanaian Chronicle* 2004). This 'national crisis' has led the National Road Safety Commission (NRSC) and other state agencies and NGOs to embark upon road safety campaigns, flanked by heated debates about who and what to blame for the crisis. At the centre of the debates was the AKR due to its 'notoriety for gruesome accidents owing not only to its poor physical nature, but also its improper design' (*Ghanaweb* n.d.). These findings were confirmed in a road traffic study conducted to identify black spots on Ghana's entire road infrastructure in 2005 following pressure to identify measures 'to limit the carnage' (*Ghanaweb* 2003b). One of the (unsurprising) findings of the study, in which forty 'death spots' were identified (*Daily Graphic* 2005), was that Potroase was deadly dangerous.[33]

According to commentators, the Potroase section had recorded the highest number of accidents in the entire Eastern Region in 2004 because of its 'improper design' – its tricky course in a particularly challenging physical environment. About sixty km beyond Accra and ten km before reaching Kyebi, the road gently climbs up into the thickly forested Atiwa mountains before suddenly plunging into a curved valley and climbing a steep slope at the other end again. At the foot of the valley lies Potroase. Where vehicles used to crash, literally falling into the village on both sides of the bend, road signs still warn motorists of the 'treacherous curve' and of 'death on the hairpin bend around the cliff' (*Ghanaweb* 2003a). In order to make the 'Potroase problem [...] a thing of the past' (*Ghanaweb* n.d.), the road agencies – and allegedly the Okyenhene too – opted for the radical engineering solution of redirecting the road. The Apedwa-Bunso bypass was built to 'make [the AKR] safe', as the Minister of Transport assured (*Ghanaweb* 2002).

32 Apart from the AKR, the coastal roads (Accra-Aflao, Accra-Cape Coast) are frequently labelled 'death traps'. Ghanaians also use the term 'death trap' as a label for trotros, taxis and other types of passenger vehicles (Hart 2013: 382).

33 See National Road Safety Commission (2006) for figures on black spots.

Dangerous Drivers, Speeding Strangers and Carelessness

People blamed the numerous Potroase accidents on the road's bad state and design, but also on the negligence of drivers. 'Potroase was a death zone because of careless driving', one Kyebi friend told me. His view resonates with the general assessment of Ghanaian road safety experts. Drawing from various (not specified) studies, they assert that 'human error (70–93%) is the most cited contributory factor underlying the occurrences of road traffic crashes' (National Road Safety Commission 2009: 13). During road safety campaigns that I attended and in conversations with representatives of the NRSC, the experts were keen on repeating the precise 93% figure. This was done out of conviction, and to convince people that safe roads were ultimately the responsibility of their users, meaning drivers. Responsible driving was advocated as the only alternative to the high percentage of human error that included 'unacceptable' road user behaviour such as impatience, recklessness and non-compliance. Experts reiterated during campaigns and in interviews that these were all easily avoidable by simply 'doing the right thing'. The media spread the same attitude. In a feature entitled 'Death on our roads', one journalist lamented that 'our roads have turned into death traps; our vehicular facility has turned into moving coffins because our drivers continue to be reckless and careless. Such is our concern' (*Ghanaian Chronicle* 2004).

By placing the responsibility of drivers at the heart of road safety debates (see Lamont 2010, 2013) and thus perpetuating the 'largely behaviorist model' (Morris 2010: 601) of educational initiatives, experts and the public tend largely to neglect – but still occasionally evoke – the complementary causes for accidents. These causes can be found on a physical-technical level (e.g. bad roads, bad vehicles) and in a legal-structural context (inadequate traffic law enforcement, bribery). They also include inexperience, lack of driving skills, economic pressures and competition among drivers, all of which contribute to the often-cited carelessness.

With regards to the Potroase accidents, my interlocutors from Kyebi and Suhum – many of whom are commercial drivers – equally stressed careless driving as a core factor. They reasoned that drivers (in particular inexperienced, foreign ones) used to employ excessive speed at Potroase. Some avoided applying their brakes when descending into the valley in order not to lose momentum – from fear that their heavy, often overloaded and weak lorries would otherwise struggle ascending again on the other side. Others simply misjudged their speed or were impatient, to save time and make profit. 'Overspeeding' (*kɔ ntɛmntɛm*, to go very fast, also: *kɔ speed*) was believed to have caused many big vehicles to fail to negotiate the sharp bend, subsequently crashing in the valley and spilling their freight onto the lower township, where houses were sometimes destroyed and residents killed.

The issue of excessive speed on the AKR was regularly raised by road safety experts. For instance, one study conducted on the AKR revealed particularly astonishing speed figures (Derry et al. 2007): Vehicles plying two road sections with a prescribed speed limit of 50 km/h were measured travelling at an average of 87 km/h, within a range of 40 to 187 km/h. 95% of all vehicles were exceeding the prescribed speed limit. The report concluded that the excessive speeds, particularly when coupled with wide speed *variations*, contributed to the high incidence of traffic crashes and fatalities on the AKR. Other statistical figures used by experts showed that more than half of all accidents on the AKR were related to overspeeding (National Road Safety Commission 2003; Fig. 4.3) or, as cited by *Ghanaweb* (2005b), to 'excessive over speeding [sic]'.[34] One central finding from my research with commercial drivers was that it is the occurrence of not only excessive, but of *differential* speeds that form a major challenge in the everyday work of these road professionals.

At Potroase, overspeeding and fatalities have eventually become 'a thing of the past' as nowadays few vehicles pass through the village. 'You see, this place is no more dangerous', the relieved residents told me and pointed to the children sitting and chatting undisturbed on the dented railguards – remnants of an attempt to prevent vehicles from crashing into the township. People have

FIGURE 4.3 *'Overspeeding KILLS', message by the* NRSC
SOURCE: G. KLAEGER, 2007

34 See Morris (2010: 601) on speeding as the primary explanatory factor for accidents in South Africa.

also observed that not only has traffic been diverted, but so has dangerous driving: soon after its opening, a series of fatal accidents started to occur on the new Apedwa-Bunso section. 'Now they go even faster because the road there is so smooth', people told me, furthering that the new, smooth road proves particularly fatal when it gets wet and 'slippery'. The observation that drivers were now taking 'undue advantage of the good nature of the road to over speed [sic], thus causing accidents' (*Ghanaian Chronicle* 2006a), as one traffic police commander explained, was also made for other newly-built sections of the AKR.

Observing how careless driving has shifted from the old to the new road, my Kyebi acquaintances found it ironic that new roads, built with the intention to reduce accidents on the AKR, continue to be a worrying source of danger. They also observed that the speed-up of the country's development, fostered by an improved road infrastructure, in turn fosters overspeeding. The correlation between nation-wide acceleration and increased, at times fatal, vehicular speed is by no means solely a Ghanaian experience (see Morris 2010 and Lee 2012 on South Africa; Lamont 2012, 2013 on Kenya). Neither is it limited to the African continent, as Árnason et al. (2007) provide a striking ethnographic example from present-day Iceland. Iceland's recent experience of modernity, neo-liberalism and accelerated development has also been one of increased movement and faster driving – an experience that 'open[s] up a road that can lead to destruction' (Árnason et al. 2007: 209). This corresponds to Paul Virilio's (1986) theoretical idea of the 'dromocratic condition' and to his assertion that acceleration is crucial to modernity, yet also that ever-increasing speed necessarily results in various types of accidents.

Powers and Spirits: Roaming Dangers in Potroase and beyond

Accidents at Potroase – and on the AKR road in general – were seen by many in Kyebi to be related to drivers' carelessness, particularly to overspeeding. Concurrently, and often without sensing any contradiction, some brought forward a further cause for the accidents at the notorious black spots: spirits and witchcraft. They argued that the presence of malicious powers on and alongside the road was undeniable in view of the suspiciously high number of accidents at some spots. Speculations about the involvement of such powers were also nurtured by the views of crashed vehicles – of broken-down trucks in odd, almost impossible positions left behind by the road, or of incredibly mangled carcasses deposited at police stations. The 'Airport' tale recounted earlier is one of many popular depictions of the road's spiritual dangers that are framed by rumours alleged to be very embarrassing to the Okyenhene. They speak of witchcraft that roadside residents have carried out in order for vehicles, especially cargo trucks, to crash, overturn and spill their goods that are then appropriated by residents approaching the accident scene. Potroase/'Airport' has regularly been referred to as a 'witch pot' as a result. Other rumours brought up not only for Potroase refer to bloodthirsty spirits that want to 'suck' the blood of accident victims. Such spirits are involved in accidents that occur particularly during festive seasons, when everybody – including spirits and souls – wants to 'enjoy something'; or else, during election times, when politicians are suspected to turn to spiritual powers that require human blood for their protective and supportive actions (see Meyer 1998).[35]

While some powers may occur and strike at any place and road section (without particular reference of them being mobile and roaming), others are said to be residential and inhabiting distinct spots and road sections. Apparently, one such locale of spiritual dangers for travellers is Asuboi, a small roadside community in the Akyem area. Much attention has been paid to the short and narrow Asuboi bridge that leads over a small creek. Over the years, vehicles that did not quite make it through the narrow road segment have badly damaged the bridge's cement railings and some have even ended up in the creek or its banks after colliding with oncoming vehicles. My friends in Kyebi had no doubts that these accidents were connected to the drivers' failures to negotiate this tricky road section. Nevertheless, they brought up the 'story' they have been told about the bridge, that the Asuboi river is allegedly the home of a deity whose children come out in the dark to play on the bridge. People told me laughingly that drivers were advised to blow their horn when crossing, not only to scare the playing children away, but also to greet the spirit. 'You have to give the spirit the appropriate respect', a friend said, suggesting that failing to do so might result in an accident.

The frequent, sometimes fatal accidents on the bumpy and hilly road section in the Asuboi area have given rise to the popular belief, or at least narrative, of the perilous presence of the spirits (or souls) belonging to those who have died there in previous accidents. 'At such places, they want more dead', people claim, meaning that the souls are eager to find other souls to join them. Similar accounts of occult forces or 'evil creatures roaming the highways' (Masquelier 1992: 60) are reported from other parts of Ghana (Verrips and Meyer 2001) and other African countries.[36] For instance, Morris (2010: 596) claims for South African contexts that 'the road is frequently narrated [...] as the space of particular vulnerability to witchcraft'. Particularly rich is the account of road stories that Masquelier (2002) collected among residents living close to Niger's Route Nationale 1. The stories reveal the presence of, and people's tragic encounters with, bloodthirsty spirits and other 'sinister forces'

35 See Blunt (2004) on the linkage between road accidents and elections in Kenya.

36 See, for example, Chilson (1999), Geschiere (1997), Giles-Vernick (1996), Lee (2012), Masquelier (1992, 2002), Sanders (2008) and Smith (2004).

(Masquelier 2002: 841). Certain accident-ridden road sections have earned a reputation as particularly dangerous places harbouring 'frightening road spirits' (Masquelier 2002: 844). According to Masquelier (2002: 845), the residents' (as well as the travellers') fears are to be seen 'as an important dimension of the imaginative practices through which people read the landscape, invest it with moral significance, and assess their personal vulnerability'.

The occult and destructive road is not restricted to a matter of creative imageries and narratives. It encourages some road users in Ghana and elsewhere to employ certain preventive and protective measures to ensure their safety on the road (see Klaeger 2009). Roadside residents equally engage in concrete actions, such as the performance of rituals at accident spots in places like Asuboi, to 'drive away' the ghosts of accident victims. Residents in Potroase recall that the late King Kuntunkununku II was once involved in a cleansing ritual at the Potroase bend as well.[37] His successor, the current Okyenhene, is alleged by some to have been in favour of a rather secular measure, namely the building of the new bypass. Such a measure not merely avoids continuous accidents, but more importantly makes it less likely that – as an educated and 'modern' traditional ruler – he would get himself involved in such ritual matters.[38]

Conclusion

At the centre of this chapter were stories and rumours surrounding a new road section recently constructed on the Accra-Kumasi road, and the various actors allegedly caught up in the construction process. These narratives provide insights into the ways in which the road is perceived as embedded in – and at times channelling, even provoking – issues of power, progress and perils. I have shown how these issues have been and continue to be of immediate concern for the roadside community of Kyebi, its residents and politicians. For them, the road is clearly a two-directional phenomenon, leading both forward and backward. What is reflected in people's concern is a sense of modernity, but also one of uncertainty, instability and threat. The latter is nurtured by

37 The only ritual road cleansing that I witnessed during my fieldwork was performed by the chiefs of Ofankor, a suburb of Accra, in late 2006 in order to appease *Olila*, the local road deity; see Klaeger (2013a) for a full account of the ritual, the involved actors and the surrounding debates.

38 The logic behind this argument is that the King is known to actively avoid involvement in such 'traditional' practices (customs, *ammamere*) in public. I discuss the Okyenhene's politics of negotiating his 'traditional' duties elsewhere (Klaeger 2007b).

their tangible experiences of repeatedly losing control over 'their' road; of being sidelined from progress, national development and, most visibly, the main through traffic, which now fosters the town's dullness; and of being confronted with death and destruction. Here the provenance and locations of threats are varied, as is the case with the centres of power and politics that are located both in Kumasi and Accra, and with passing vehicles in contrast to stationary malicious forces. The threats are intrinsically connected with the AKR – whether it worked as a contested trade route that served as a path of authority, or emerges as a feared death trap in which travellers, sometimes residents as well, get killed.

Yet there remain further threats for this roadside community, namely from those that come by the road and leave on it again after having perpetrated what Kyebi's residents fear, and vividly debate, perhaps more than the above. It is the coming and going of criminals, mostly at night, such as the thieves who stole goats from a compound near the main road with their pickup truck in which they 'ran away' again; or the armed robbers who broke into a household in town in search of valuables and who shot and killed the landlord before escaping via the road by which they had allegedly come. Clearly, for Kyebi's residents, the experience of the road has been often, and recurrently, a negative and sobering one. Yet it is also worth noting that this experience is accompanied by people's engagement with the road as a place and path imbued with possibilities and promises, some of which are realised and fulfilled on a regular, even routine basis. This experience and engagement forms the ground for the often-noted ambivalent stance that people take towards roads (see Klaeger 2013b; Masquelier 2002) – an ambivalence that crucially contributes to the everyday making of 'their' roads.

References

Accra Daily Mail. 2006. 'President Kufuor Inaugurates Ofankor-Nsawam Dual Carriageway' (20 June). Accessed November 24, 2006. http://www.accra-mail.com/mailnews.asp?ID=16999.

Affrifah, Kofi. 2000. *The Akyem Factor in Ghana's History, 1700–1875*. Accra: Ghana Universities Press.

Alber, Erdmute. 2000. 'Automobilismus und Kolonialherrschaft: Zur Bedeutung des Autoverkehrs für die Herrschaftsstrukturen in der westafrikanischen Kolonie Dahomey.' *Paideuma* 46: 279–299.

Alber, Erdmute. 2002. 'Motorization and Colonial Rule: Two Scandals in Dahomey, 1916.' *Journal of African Cultural Studies* 15 (1): 79–92.

Árnason, Arnar, Sigurjón Baldur Hafsteinsson and Tinna Grétarsdóttir. 2007. 'Acceleration Nation: An Investigation into the Violence of Speed and the Uses of Accidents in Iceland.' *Culture, Theory and Critique* 48 (2): 199–217.

Blunt, Robert. 2004. '"Satan is an Imitator": Kenya's Recent Cosmology of Corruption.' In *Producing African Futures: Ritual and Reproduction in a Neoliberal Age*, edited by Brad Weiss, 295–328. Leiden: Brill.

Boni, Stefano. 1999. 'Striving for Resources or Connecting People? Transportation in Sefwi (Ghana).' *The International Journal of African Historical Studies* 32 (1): 49–70.

Bowdich, Thomas Edward. 1819. *Mission from Cape Coast Castle to Ashantee, with a Statistical Account of that Kingdom, and Geographical Notices of Other Parts of the Interior of Africa*. London: J. Murray.

Campbell, Jeremy M. 2012. 'Between the Material and the Figural Road: The Incompleteness of Colonial Geographies in Amazonia.' *Mobilities* 7 (4): 481–500.

Chilson, Peter. 1999. *Riding the Demon. On the Road in West Africa*. Athens and London: University of Georgia Press.

Christaller, Johann Gottlieb. 1881. *A Dictionary of the Asante and Fante Language Called Tshi (Chwee, Twi)*. Basel: Basler Mission.

Colombijn, Freek (ed.). 2002. *On the Road: The Social Impact of New Roads in Southeast Asia*. Leiden: KITLV.

COMPTRAN n.d. 'Transportation/Higway Projects.' Accessed March 30, 2011. http://www.comptran-eng.com/projects/ksi_highway.html.

Daily Graphic. 2005. '40 Death Spots Identified' (25 November). Accessed May 23, 2013. http://www.ghanaweb.com/GhanaHomePage/NewsArchive/artikel.php?ID=95006.

Dalakoglou, Dimitris. 2010. 'The Road: An Ethnography of the Albanian-Greek Cross-Border Motorway.' *American Ethnologist* 73 (1): 132–149.

Derry, James, Francis Afukaar, Peter Donkor and Charles Mock. 2007. 'Study of Vehicle Speeds on a Major Highway in Ghana: Implication for Monitoring and Control.' *Traffic Injury Prevention* 8 (2): 142–146.

Dickson, Kwamina B. 1961. 'The Development of Road Transport in Southern Ghana and Ashanti since about 1850.' *Transactions of the Historical Society of Ghana* 5 (1): 33–42.

Drummond-Thompson, Philip. 1992. 'The Development of Motor Transport in Nigeria: A Study in Indigenous Enterprise.' In *Les Transports en Afrique (XIXe–XXe siecle)*, edited by Helene d'Almeida-Topor, Chantal Chason-Jabeur and Monique Lakroum, 222–247. Paris: Editions L'Harmattan.

Dupuis, Joseph. 1824. *Journal of a Residence in Ashantee*: London.

Fairhead, James. 1993. 'Paths of Authority: Roads, the State and the Market in Eastern Zaire.' In *Real markets: social and political issues of food policy reform*, edited by Cynthia Hewitt de Alcántara, 17–35. London: Frank Cass.

Fynn, John Kofi. 1973. 'Asante and Akyem Relations 1700–1831.' *Institute of African Studies Research Review* 9 (1): 58–81.

Geschiere, Peter 1997. *The Modernity of Witchcraft: Politics and the Occult in Postcolonial Africa.* Charlottesville: University of Virginia Press.

Gewald, Jan-Bart. 2002. 'Missionaries, Hereros, and Motorcars: Mobility and the Impact of Motor Vehicles in Namibia before 1940.' *The International Journal of African Historical Studies* 35 (2/3): 257–285.

Gewald, Jan-Bart, Sabine Luning and Klaas van Walraven. 2009. 'Motor Vehicles and People in Africa: An Introduction.' In *The Speed of Change. Motor Vehicles and People in Africa, 1890–2000*, edited by Jan-Bart Gewald, Sabine Luning and Klaas van Walraven, 1–18. Leiden: Brill.

Ghanaian Chronicle. 2004. 'More Deaths Result From Road Accidents Than HIV/Aids, Say GRSC Chairman' (15 April). Accessed May 23, 2013. http://allafrica.com/stories/200404150565.html.

Ghanaian Chronicle. 2006a. 'Kyebi Records Highest Accident Rate in E/Region' (23 January). Accessed January 31, 2006. http://allafrica.com/stories/200601231031.html.

Ghanaian Chronicle. 2006b. 'Mercedes Benz 207, Ford Express And Road Accidents in Ghana' (23 March). Accessed May 23, 2013. http://allafrica.com/stories/200603230493.html.

Ghanaweb. 2002. 'Gov't Awards Road Construction Contract to Ivorian Company' (25 April). Accessed January 18, 2006. http://www.ghanaweb.com/GhanaHomePage/NewsArchive/artikel.php?ID=23569.

Ghanaweb. 2003a. 'Elegy From The Bowels Of Kwaebibirem' (28 February). Accessed January 18, 2006. http://www.ghanaweb.com/GhanaHomePage/features/artikel.php?ID=33399.

Ghanaweb. 2003b. 'IT IS TIME: Reducing Road Accidents on Ghanaian Roads' (28 October). Accessed May 23, 2013. http://www.ghanaweb.com/GhanaHomePage/features/artikel.php?ID=45612 (accessed 23 May 2013).

Ghanaweb. 2005a. 'Ghana could be Built through Serious Evangelism' (5 April). Accessed January 12, 2006. http://www.ghanaweb.com/GhanaHomePage/NewsArchive/artikel.php?ID=89496.

Ghanaweb. 2005b. 'GPRTU Official Enumerates Causes of Accidents' (16 December). Accessed January 17, 2006. http://www.ghanaweb.com/GhanaHomePage/NewsArchive/artikel.php?ID=96190.

Ghanaweb. n.d. 'The Suhum "Bermuda Triangle"'. Accessed February 1, 2006. http://www.ghanaweb.com/concord/article.php?ID=403.

Giles-Vernick, Tamara. 1996. 'Na lege ti guiriri (On the Road of History): Mapping Out the Past and Present in M'Bres Region, Central African Republic.' *Ethnohistory* 43 (2): 225–275.

Gould, Peter R. 1960. *The Development of the Transport Pattern in Ghana*. Evanston, Illinois: Northwestern University.

Hart, Jennifer. 2013. '"One Man, No Chop": Licit Wealth, Good Citizens, and the Criminalization of Drivers in Postcolonial Ghana.' *International Journal of African Historical Studies* 46 (3): 373–396.

Harvey, Penelope. 2005. 'The Materiality of State-Effects: An Ethnography of a Road in the Peruvian Andes.' In *State Formation: Anthropological Perspectives*, edited by Christian Krohn-Hansen and Knut G. Nustad, 216–247. London, Ann Arbor, MI: Pluto Press.

Harvey, Penelope. 2012. 'The Topological Quality of Infrastructural Relation: An Ethnographic Approach.' *Theory, Culture and Society* 29 (4–5): 76–92.

Harvey, Penny and Hannah Knox. 2012. 'The Enchantments of Infrastructure.' *Mobilities* 7 (4): 521–536.

Havik, Philip J. 2009. 'Motorcars and Modernity: Pining for Progress in Portuguese Guinea, 1915–1945.' In *The Speed of Change. Motor Vehicles and People in Africa, 1890–2000*, edited by Jan-Bart Gewald, Sabine Luning and Klaas van Walraven, 48–74. Leiden: Brill.

Heap, Simon. 1990. 'The Development of Motor Transport in the Gold Coast, 1900–39.' *Journal of Transport History* 11 (2): 19–37.

Huppenbauer, David. 1888. *Von Kyebi nach Kumase: Eine Reise ins Hinterland der Goldküste*. Basel.

Ispahani, Mahnaz Z. 1989. *Roads and Rivals: The Political Uses of Access in the Borderlands of Asia*. Ithaca: Cornell University Press.

Jockers, Heinz, Dirk Kohnert and Paul Nugent. 2010. 'The Successful Ghana Election of 2008: A Convenient Myth?' *Journal of Modern African Studies* 48 (1): 95–115.

Khan, Naveeda. 2006. 'Flaws in the Flow: Roads and Their Modernity in Pakistan.' *Social Text* 24 (4): 87–113.

Klaeger, Gabriel. 2007a. *Kirche und Königspalast im Konflikt; Christenviertel und Tradition in den Debatten zum Ohum-Festival in Kyebi (Ghana)*. Hamburg: LIT.

Klaeger, Gabriel. 2007b. 'Palastbräuche versus Missionarsrechte: Traditionspluralismus und angefochtene Souveränitäten im Konflikt um das Ohum-Festival in Kyebi (Ghana).' *Zeitschrift für Ethnologie* 132 (1): 91–108.

Klaeger, Gabriel. 2009. 'Religion on the Road: The Spiritual Experience of Road Travel in Ghana.' In *The Speed of Change. Motor Vehicles and People in Africa, 1890–2000*, edited by Jan-Bart Gewald, Sabine Luning and Klaas van Walraven, 212–231. Leiden: Brill.

Klaeger, Gabriel. 2012. 'Rush and Relax: The Rhythms and Speeds of Touting Perishable Products on a Ghanaian Roadside.' *Mobilities* 7 (4): 537–554.

Klaeger, Gabriel. 2013a. 'Dwelling on the Road: Routines, Rituals and Roadblocks in Southern Ghana.' *Africa* 83 (3): 446–469.

Klaeger, Gabriel. 2013b. 'Introduction: The Perils and Possibilities of African Roads.' *Africa* 83 (3): 359–366.

Lamont, Mark. 2010. 'An Epidemic on Wheels? Road Safety, Public Health and Injury Politics in Africa.' *Anthropology Today* 26 (5): 3–7.

Lamont, Mark. 2012. 'Accidents Have No Cure! Road Death as Industrial Catastrophe in Eastern Africa.' *African Studies* 71 (2): 174–194.

Lamont, Mark. 2013. 'Speed Governors: Road Safety and Infrastructural Overload in Post-Colonial Kenya, c. 1963–2013.' *Africa* 83 (3): 367–384.

Lee, Rebekah. 2012. 'Death in Slow Motion: Funerals, Ritual Practice and Road Danger in South Africa.' *African Studies* 71 (2): 195–211.

Masquelier, Adeline. 1992. 'Encounter with a Road Siren: Machines, Bodies and Commodities in the Imagination of a Mawri Healer.' *Visual Anthropology Review* 8 (1): 56–69.

Masquelier, Adeline. 2002. 'Road Mythographies: Space, Mobility, and the Historical Imagination in Postcolonial Niger.' *American Ethnologist* 29 (4): 829–855.

McCaskie, Tom C. 2001. *Asante Identities: History and Modernity in an African Village, 1850–1950*. Edinburgh: Edinburgh University Press.

Meyer, Birgit. 1998. 'The Power of Money: Politics, Occult Forces, and Pentecostalism in Ghana.' *African Studies Review* 41 (3): 15–37.

Ministry of Roads and Transport. 2003. *Road Sector Development Programme: Review Report*. Government of Ghana.

Ministry of Roads and Transport. 2004. *Road Sector Development Programme: Programme Management Report*. Government of Ghana.

Ministry of Transportation. 2007. *Road Sector Development Programme: Review Report*. Accra: Government of Ghana.

Moriarty, Patrick and Clive S. Beed. 1989. 'Transport in Tropical Africa.' *Journal of Modern African Studies* 27 (1): 125–132.

Morris, Rosalind C. 2010. 'Accidental Histories, Post-Historical Practice? Re-Reading Body of Power, Spirit of Resistance in the Actuarial Age.' *Anthropological Quarterly* 83 (3): 581–624.

National Road Safety Commission. 2003. *Annual Report*. Government of Ghana.

National Road Safety Commission. 2006. *Evaluation Report on the 2001–2005 National Road Safety Strategy*. Government of Ghana.

National Road Safety Commission. 2009. *National Road Safety Policy*. Government of Ghana.

Nugent, Paul. 2001a. 'Ethnicity as an Explanatory Factor in the Ghana 2000 Elections.' *African Issues* 29: 2–7.

Nugent, Paul. 2001b. 'Winners, Losers and also Rans: Money, Moral Authority and Voting Patterns in the Ghana 2000 Election.' *African Affairs* 100 (403): 405–428.

Rathbone, Richard. 1989. 'A Murder in the Colonial Gold Coast: Law and Politics in the 1940s.' *The Journal of African History* 30 (3): 445–461.

Rathbone, Richard. 1993. *Murder and Politics in Colonial Ghana*. New Haven & London: Yale University Press.

Rathbone, Richard. 2000. *Nkrumah and the Chiefs*. Oxford.

Sanders, Todd. 2008. 'Buses in Bongoland: Seductive Analytics and the Occult.' *Anthropological Theory* 8 (2): 107–132.

Smith, James H. 2004. 'Of Spirit Possession and Structural Adjustment Programs: Government Downsizing, Education, and their Enchantments in Neoliberal Kenya.' In *Producing African futures: ritual and reproduction in a neoliberal age*, edited by Brad Weiss, 262–293. Leiden: Brill.

Stasik, Michael. 2013. 'In the Hustle Park: The Social Organization of Disorder in a West African Travel Hub.' *Working papers of the SPP 1448*, Halle and Leipzig, Nr. 1.

Urry, John. 2007. *Mobilities*. Cambridge: Polity Press.

van der Laan, H. Laurens. 1981. 'Modern Inland Transport and the European Trading Firms in Colonial West Africa.' *Cahiers d'Etudes africaines* 21 (84): 547–575.

Verrips, Jojada, and Birgit Meyer. 2001. 'Kwaku's Car: The Struggles and Stories of a Ghanaian Long-Distance Taxi-Driver.' In *Car Cultures*, edited by Daniel Miller, 153–184. Oxford: Berg.

Virilio, Paul. 1986. *Speed and Politics*. New York: Semiotext(e).

Wilks, Ivor. 1975. *Asante in the Nineteenth Century: The Structure and Evolution of a Political Order*. London: Cambridge University Press.

Wilks, Ivor. 1992. 'On Mentally Mapping Greater Asante: A Study of Time and Motion'. *The Journal of African History* 33 (2): 150–190.

Wilson, Fiona. 2004. 'Towards a Political Economy of Roads: Experiences from Peru.' *Development and Change* 35 (3): 525–546.

World Bank. 2008. *Ghana – Road Sector Development Project*. Washington D.C.: The World Bank. Accessed May 23, 2013. http://documents.worldbank.org/curated/en/2008/12/10276477/ghana-road-sector-development-project.

Wrangham, Elizabeth. 2004. 'An African Road Revolution: The Gold Coast in the Period of the Great War.' *The Journal of Imperial and Commonwealth History* 32 (1): 1–18.

Yarak, Larry W. 1990. *Asante and the Dutch: 1744–1873*. Oxford: Clarendon.

CHAPTER 5

Biographies of Roads, Biographies of Nations

History, Territory and the Road Effect in Post-conflict Somaliland

Luca Ciabarri[1]

The road that connects the port of Berbera in the Gulf of Aden to the Ethiopian border and which continues to the Ethiopian highlands was at the heart of the process of formation and consolidation of the neo-state of Somaliland.[2] It is, indeed, no exaggeration to argue that this road 'made' the state. Representing a major axis along which the new polity took shape and new institutions were formed out of the Somali civil war in the early 1990s, this road has since entailed many of the hopes, realities and failures of the country's development. The aim of this chapter is to qualify this assertion by developing an ethnographic analysis of the road and to reflect upon the possibilities and limitations of such an approach. The situation of institutional vacuum that was initially created by the Somali war has highlighted specific features of the relationship between the road – conceived as a potential for mobility and connection – the territory it came to occupy, and its inhabitants. Before analysing the ethnographic data however, this chapter begins by providing a few methodological and historical remarks.

1 The chapter draws on fieldwork carried out between 2007 and 2008 along Somaliland roads. Fieldwork and writing have been possible thanks to substantial support from the Max Planck Institute for Social Anthropology, Halle-Saale, Germany, where I worked from 2006 to 2009 and in 2011 from the Wenner-Gren Foundation (Hunt fellowship). Earlier versions of the article were presented in conferences in Djibouti, Leipzig and Thurnau. I'd like to thank the participants of these gatherings for their invaluable feedback, specifically the members of the 'African roads group' convened in Thurnau 2012 and Lisbon 2013 workshops under the organisation of Kurt Beck, Gabriel Klaeger and Michael Stasik.

2 Somaliland, internationally not recognised, is a de-facto state, occupying the territory of the former British Protectorate. It emerged in 1991 from the dissolution of Somalia after three years of internal struggle with the central regime.

 | DOI 10.1163/9789004339040_006

The State and the Road, or Who Builds What along the Road: Ethnographic Approaches

In recent anthropological studies, roads have frequently been associated with the colonial or postcolonial state and with international institutions involved in development and nation building projects. Ethnographic accounts (Dalakoglou 2010; Harvey 2005; Masquelier 2002; Pina-Cabral 1987) have described roads as expressing a set of official meanings (modernity, national integration, development, future) whilst simultaneously showing the multiple ways in which local actors appropriate roads (how they are reworked practically and symbolically and, at times, even rejected, often in the very process of their construction). Roads thus also come to be linked to creativity, change, ambiguity, risk and danger, as well as to the threat of losing control and to external influences. In this respect, the case that is presented in this chapter is slightly different and it helps to focus on other facets of roads. The Somaliland case does not fall into the kind of narrative of 'built by the state, the road represents… (modernity, development, speed, connection, threat, local tensions etc.)'. Rather it can be conceived by the hypothetical narrative of 'built by the road, the state represents…', which simply hints at the effects generated by a road in its relationship with the territory and its inhabitants, including political effects.

Alongside its material configurations (paved road, gravel road etc.), a road is also a form of territorial integration that requires, among others, a mode of regulation of transit and movement. The road is therefore not only a physical construction but also a social and political one, which requires a number of preconditions to ensure that things and people can actually move along it. This became all the more evident in a situation of institutional breakdown, like in Somalia during the civil war. The processes of reconstruction and reformulation of these conditions have basically corresponded to the processes of the formation of Somaliland. By integrating the road – alongside the port and the border zone – into a specific economic and political project, these dynamics gave rise both to a transit economy as well as to a commercial state. Territorial and historical dimensions also play a role: as a form of territorial organisation, the road has represented a site of opportunity in times of crisis, dictating specific patterns of economic and political behavior and articulating specific geopolitical designs deeply rooted in local social memories. In the long run, roads do not simply adapt to their territory but produce and refashion it too.

In order to further clarify this point, we should first highlight the simple fact that there is no single point of view nor is there a single focus for a road – as much as there is not one single direction. A road can therefore never be observed in its entirety, and any specific point on a road basically implies a 'somewhere else'. This also means that a road is external and extraneous to any local context, despite it being intricately interwoven with it. Incidentally, this is why roads are frequently perceived as something ambiguous, as a threat or an adventure. A sense of totality of a road can only be constructed and imagined by different social actors aiming at different objectives.[3] Generally, this totality refers to the potential of a road for connection and mobility; but this potential is then realised in many different ways and is directed towards multiple goals.

It is through this constructed sense of totality (along the axes of mobility and connection) that the road de facto produces social effects, to paraphrase Mitchell's (1999) notion of state effect.[4] In imagining a form of totality of a road, forcefully partial and subjective, social actors interweave it with their own strategies, plans, actions, hopes, imaginations and tools. The road effect lies in this interweaving, in which a road is no longer an inert infrastructure but a social and political construction, managing, producing or performing mobility and connection.

My main interest for this chapter pertains to the long-term effects of the regional commercial corridors in the northeast Horn of Africa. This set of effects results from a number of unrelated actions and actors. Consequently, the ethnographic representation has to take into account all these factors as well as the entanglements of road stories with various forms of local social histories, constructing a sort of 'biography of the road'. In the case study under consideration, this includes accounts on trade, movement of people, war and peace, transport, the knots along the road in the form of urban centres, the port as well as the border zone.[5] In order to account for the different totalities built through the road, we have to intellectually move along it, and consequently connect one point with the other in writing the road.

3 This is true for any phenomenon where presence and absence are simultaneously implicated. As for the experience of regionality, Nancy Munn (1990: 2) similarly argues that 'for a subject, a regional world is not given but lived and created' through 'experiential synthesis' and 'spatiotemporal connectors', which are rooted in history and socio-cultural patterns.

4 Timothy Mitchell questions the idea of permanence and solidity of the state by arguing that this idea, rather than being consubstantial to it, is constantly constructed by social actors, producing a state effect and thus the capacity of the state itself to act (Mitchell 1999).

5 The term 'biography' refers to Kopytoff's and Appadurai's reflections (1986), but is used here in a broad sense, referring to the reconstruction of the social histories and identities of groups, individuals and places found along the road.

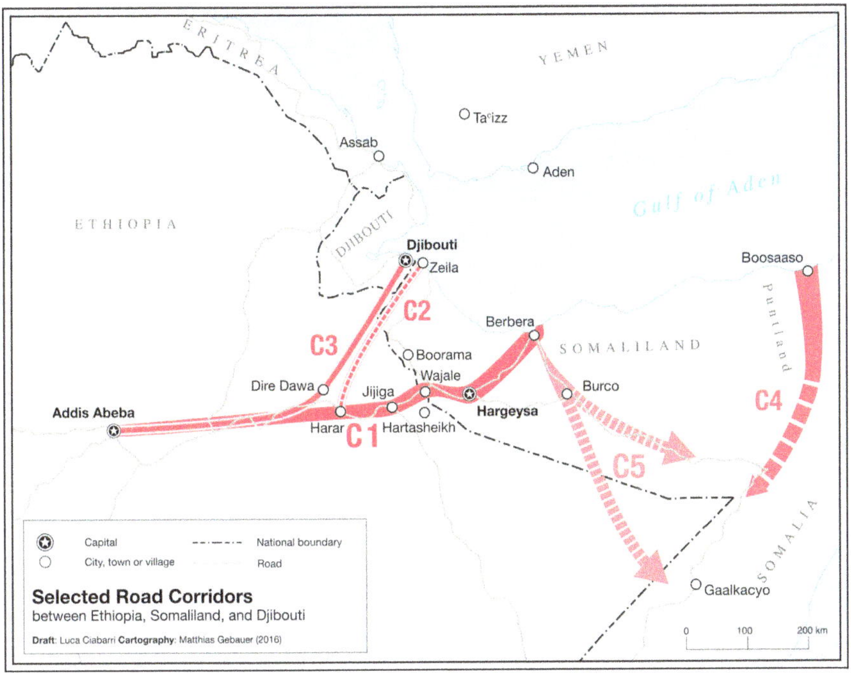

MAP 5.1 *Selected road corridors between Ethiopia, Somaliland, Djibouti*

Roads for Empires, for Development, for Trade

The Berbera corridor, as it is named in the current developmental terminology (EC Kenya 2003), runs from the port city of Berbera, Somaliland, to Addis Ababa, Ethiopia, with a total length of 937 km (C1 in Map 5.1).[6] The Somaliland stretch corresponds to 241 km, even though the Somali-inhabited area runs well inside Ethiopia, reaching the town of Jijiga and the ethnically mixed region of Harar. The stretch between Berbera and Harar represents the historical trading route. The two centres are the most ancient towns along the road, while Jijiga and the current capital city of Somaliland, Hargeysa, emerged more recently (Lewis 2002). Smaller towns and villages took shape along the road in the 20th century, which was due to the road effect, the expansion of the state and the decline of nomadic pastoralism. In addition to the Berbera corridor, the region

6 In this article, the term 'corridor' refers to the roads' function (connecting the sea to the inland, primarily for commercial purposes) that I focus on, irrespective of their particular shape and materiality. As long as the roads fulfill this function, I use terms like road corridor, commercial corridor and commercial route interchangeably.

is also crossed by other routes leading to the Ethiopian highlands. All these corridors are in direct competition with each other, but, as will be shown, a closer look reveals a number of relevant intersections.

Other corridors include, west of Berbera: (1) the historical route from the port of Zeila[7] (Somaliland) to Ethiopia (C2 in Map 5.1), which in colonial times was replaced by (2) the corridor from Djibouti to Addis Ababa (C3 in Map 5.1). Given Ethiopia's lack of accessibility to the Eritrean ports, the motor road leading through this corridor is now the principal trading route for the land-locked country. East of Berbera (3) is the Boosaso corridor[8] (C4 in Map 5.1) which has emerged in recent years as a result of the construction of a road and a port in the late 1980s. This has made Boosaso a leading centre of the new transit economy so peculiar of the 1990s and 2000s Somalia. Finally, from Berbera departs also (4) the Berbera-Burco route (C5 in Map 5.1), which links Berbera to eastern Somaliland as well as to southern Somalia and eastern Ethiopia.

The Italian Army built a proper motor road following the ancient caravan routes that connects Harar to Berbera during the invasion campaign of the British Protectorate of Somaliland in 1941. The occupation lasted only a few months before it was brought to an end by the British counteroffensive, but the road layout remained.[9] In the 1960s a Lebanese company asphalted the stretch of the road from Berbera to Hargeysa utilising European funds. The project marks the initial years of independence and attests to the importance of the port of Berbera for the new Somali state.

Asphalting the section from Hargeysa to the Ethiopian border was carried out in the late 1970s and 1980s, and proceeded very intermittently: an Italian company completed one section, then a Chinese one took over the task up to the village of Kalabeydh. Approximately 15 more kilometres were eventually realised by Somali constructors until Dilla. The border, however, was never reached and a gravel road rehabilitated by international aid in late 1990s still connects Kalabeydh to the boundary. The entire project suffered from the tense relationship between Ethiopia and Somalia, as a consequence of the 1977–78 Ogaden war fought between the two states, and later on from the growing instability within Somalia. Even today it is still possible to notice the

7 Somali: Saylac.

8 Located in the semi-autonomous region of Puntland, another entity that has emerged from the dissolution of Somalia.

9 Together with the economic and social changes brought about by the expansion of roads in the region due to these competitive imperial designs (Geshekter 1985). For the relevance of motor vehicles and traders to African social history see Gewald, Luning and van Walraven 2009.

differences between these sections of the road: despite uniform repair works, time continues to produce different potholes and damages, which are a consequence of the different materials and techniques used by the various construction companies.[10]

After the civil conflict (1988–1991), a campaign of substantial repair works was launched with EU funds assigned to post-conflict rehabilitation and local institution building (EC Kenya 2003) and a Somaliland Road Authority was established. Again, the fragility of local development and international rivalries emerged: the EU rehabilitation projects carried out in the 1990s, involving the entire Berbera corridor, remained poorly funded in the 2000s, while in Ethiopia, Chinese companies stepped in to build an asphalt road from Harar to Jijiga and from there to the international border.

Empire making, war making, development making and commercial designs have over the years thus represented different power configurations that have shaped the road. But whether they are paved or not; rehabilitated gravel roads, or pure sandy tracks, the central feature of the road network in the region is represented by the intricate set of transit corridors connecting the sea to the inland.

Competing Views of the Regional Corridors, Past and Present

Within this framework, three general factors directly support my initial claim regarding the link between the road and the emerging nation of Somaliland.

The first factor addresses history and evokes the strengths of the past – of structures of *longue durée* – in guiding current changes: the Somali coasts have historically played the role of an interface and a transit zone between Ethiopia and the sea. As mentioned above, the road from Berbera to Ethiopia corresponds to an historical trading route that has for centuries connected the Ethiopian highlands to the Somali lowlands and the sea with alternating fortunes (Burton 1987; Robecchi Bricchetti 1896; Swayne 1903).[11] This historical dimension resonates with present days and determines patterns of economic and social behaviour and of territorial specialisation. In the contemporary context, the current paved road rests on an infrastructure of memories, social

10 Information on road constructions has been collected through interviews with truck drivers and traders during fieldwork in 2007–08 – no official records are available after the civil conflict.

11 For classical and recent analysis of the importance of trade and trading routes in the region see Abir 1980; Freitag 2003; Hersi 1977; Lewis 2002; Miran 2010.

institutions and structures of power, which historically served to regulate the flow of goods and people in the region and which today continues to represent horizons of development and opportunities.

The second factor draws upon the economic developments of Somaliland after the civil war and is centred on trade. A vibrant private sector and a strong business class have been described as pivotal determinants of the reconfigurations of the Somali territories during the civil war as well as in subsequent years (Bradbury 2008; Ciabarri 2010; Little 2003; Marchal 1996).[12] As the next sections will show, Somaliland business and trade correspond, above all, with the port of Berbera. Besides the port, however, trade and business in the region also comprise roads, including the connection to Ethiopia, border crossing and travel as well as international networks, urban markets and the diaspora (Ciabarri 2010). As a result, none of these aspects can be considered individually; and when combined, they construct a techno-political complex of transit, mobility and connection in which the road – and its effects – clearly plays a crucial role. Trade, in particular long-distance trade and its local ramifications, epitomises the essence of the road and until recent years, the road has literally been a synonym for trade. More specifically, the Berbera corridor is now part of a globalised trade which connects Somaliland to Ethiopia on the one side and the markets of Dubai, South-East Asia and China on the other.

The third and final factor concerns Ethiopian national interests and the re-opening of the border between Somaliland and Ethiopia in the early 1990s which intersected with Ethiopian strategic objectives of safeguarding its access to the sea. Eritrean independence in 1991 and the 1998 Ethiopian-Eritrean border war turned Ethiopia into a land-locked country, depriving it of its habitual sea outlets, the ports of Massawa and Assab in Eritrea. The events set a specific strategic scenario across the region and a tense interplay of primary national interests, involving the entire Horn of Africa and instilling a strong sense of competition between the different seaports and commercial corridors.[13] The competition involves a diverse range of actors, such as national states, international institutions, local businessmen and other local actors as well as specific political stakes – access, control, regulation, use and appropriation of the

---

12 For a regional perspective on the redrawing of the political map of Somalia, Ethiopia and Eritrea see Schlee 2003. On spatial reconfigurations and political crises see Mbembe 2000, on spatial analysis Howard and Shain 2005.

13 From Ethiopia, possible corridors to the sea run towards Port Sudan, the Eritrean ports, Djibouti, Mogadishu, Berbera and Bosaaso in Somalia or Mombasa and the Lamu port under construction in Kenya.

corridors – and techno-political stakes, ranging from road construction to road maintenance. It is precisely within this variety of socio-political dynamics and actors that ethnography sets in.

Road Users: Truck Drivers and Traders

In 2008, Ahmed 'Serwai' was working in the urban transport sector in Awdal, a neighbouring region of the main Berbera road. He was retired as a truck driver as he had been unable to return to his previous job after the civil war. His nickname, Serwai, came from the fact that he was the first in his region to drive a ZY Hino truck. Introduced in Northern Somalia in the 1980s, the Hino ZY marked a difference from the lighter Japanese trucks of the 1970s. Hino ZYs were the new, heavy trucks that began to replace the European lorries that had been in use since the colonial and early postcolonial years. 'In the 1960s to buy a truck was relatively cheap, the Somali shilling was linked to the US dollar', Ahmed Serwai related. 'Then inflation came. Japanese trucks in the 1970s were cheaper and accessible. Japanese since the 1970s have equipped all Africa.' Serwai is well known for being one of the best drivers in this region. According to his comments, however:

> Nowadays it is no longer important to be a good driver. Normally now there is a second pilot in the cabin who belongs to the subclan of the truck owner and who sooner or later learns how to drive and takes your seat. In the past trust was more important an element in the relationships between driver and owner.

Whether it is true or not, the statement points to a common change in post conflict regulations. In the early 1990s, the combination of the owner's and driver's clan was associated with the problem of security and protection: for any single tribal territory, numerous roadblocks punctuated the road and traders used to organise themselves into convoys of trucks with armed guards. Insecurity along the road dates back to the years preceding the civil war. For example, Farah, another retired driver, spent much of our conversation describing how he was repeatedly ambushed along the road in those years. Later, when the roadblocks were dismantled, the relationships between drivers and owners continued to be confined to the clan boundaries – which was also due to the logic of the internal redistribution of resources in times of economic difficulties.

'Now', Ahmed added, 'old ZYs continue to circulate together with Nissan UDs, but the newest trucks introduced in the country are Mercedes'. One important trader, in particular, has bought an entire fleet of Mercedes trucks as part of a strategy to gain power in the import-export market by means of predominance on the road. In his own words: 'There's a great scarcity of trucks now in the country, I can't run the risk of finding no trucks when I need one.' Ahmed's view diverges, however, and alludes to widespread complaints that are shared by old truck drivers, retailers and traders. Those who had been operating before the civil war and were displaced by a new generation described as aggressive and striving to control all sectors of the market and leaving no room for smaller or independent actors: 'New big traders have appeared, they even have trucks and do not rely on agents or subcontractors, they don't distribute business to other people. With them, the market is suffocating.' Though romanticised, the comparison with the past captures important elements of the transformation. Further changes have also resulted in different maps that truck drivers have drawn on the territory with their travels. In the 1980s, Serwai could find enough work because the road to Djibouti – which crosses 'his' region – was used frequently by heavy trucks transporting goods (and, as he added, remittance money) whereas now, the road is used by 4-wheel-drive vehicles carrying people, and only rarely by trucks. Big traders tend to use the Berbera corridor. Ahmed's and Farah's complaints hint at several distinct factors of post-conflict transformations: the issue of security, the importance of Djibouti in the 1980s and its more recent relative decline; the emergence of new actors and of new forms of organisation and regulation as well as new goods traded and new monopolistic tendencies.

The Port, the Road and the Border: Destruction and Opportunities within the Techno-political Complex of Mobility and Connection

Let us now turn our attention from the truck drivers to the big local traders who operate the import-export business. It was specifically in their eyes that the post-war events could present huge opportunities. Their actions paralleled the first steps of Somaliland independence, thus the troubled years between 1991 and 1993. In fact, it was only in 1993, with the rise to power of Ibraahim Cigaal,[14] that Somaliland definitely set off on a path of real stabilisation.

14 Ibrahim Egal, without using the Somali transliteration system.

As reported by A.Y. Farah and I.M. Lewis (1997: 370) the road segment between Hargeysa and Boorama (similarly to any other Somaliland road) could have up to 14 checkpoints in 1993, manned by freelance militias who were operating on their respective sub-clan's territory. Heirs of the clan-based battalions that had formed the SNM (Somali National Movement, the north-based armed movement that fought against dictatorship in the 1980s), these militias could barely be considered as under the control of the clan groups. The groups were largely comprised of ex-combatant youths or young men (called *dey-dey*) in search of a living by ways of plundering and robbery. Their incursions along the road were perceived as the most visible sign of ongoing instability, threatening the success of the peace process and the transition of SNM from an armed opposition movement to a civil government.

During those years, the port of Berbera also witnessed huge instability. Its occupation by SNM troops in 1991 paved the way for a series of national assemblies held in Berbera and Burco (in May 1991). Eventually, the national assemblies led to the Somaliland declaration of independence and to the government of Cabdiraxman Tuur, the last SNM chairman. Tuur's administration, however, collapsed in Berbera, as the new government got involved in the so-called 'sheep war', fighting over the control of port revenues.

Apart from instabilities, the effects of the war, the collapse of the central Somali state and the regional independence also brought about profit-making prospects. Berbera is again the best site from which to consider this statement. As an ex SNM combatant and mayor of the city in the early 1990s stated:

> Like any war, destruction creates opportunities, if one is ready to take some risk. This is how things started. Listen, Berbera has always been crucial for any Somali state: for the revenues from import and export, for the strategic position of its port and airstrip, and because ships from Berbera to Saudi Arabia take 48 hours, while from Mogadishu they take about a week.[15]

Opportunities not only arose from the central state's loss of custody over the port – the whole regional scenario was changing. From an economic

15 Interview, February 2008 – this is not secret information, rather a form of local conventional wisdom. The port of Berbera has always represented a major asset for the Somali state since the times of independence, and, before that, for the British Protectorate. Since the 1960s, channelling most of the booming livestock exports towards the Arab peninsula, it became one of the most important sources of income for the state, particularly in terms of custom fees and foreign currency (Aronson 1980; Samatar 1989).

viewpoint, the institutional collapse in the new Somaliland disclosed the potential of a new internal market – one that was not subjected to regulations or taxation; or to competition from external actors, but which had privileged access to the much larger and remunerative Ethiopian market via the Berbera route. The effects of state collapse in Somalia then coincided with changes in the upper part of the Berbera corridor in Ethiopia. Here, the fall of President Mengistu in May 1991 led to the reopening of the border with Somalia, which had been closed and highly patrolled since the 1977–78 Ogaden war. In addition, the turning point of 1989–90 had also important effects in Djibouti, a place that has been pivotal for regional trade in the previous decade. Here internal struggles over power sharing reached a peak and nearly led to a violent civil war, which made business practice inside Djibouti precarious and induced the Djibouti based businessmen to carefully follow the events in the neighbouring countries.

As confirmed in interviews with local traders, this vast regional scenario represents the 'natural environment' wherein local big traders who operate along the import-export chain normally situate their plans and actions.[16] The changes described above affected their ability to transport goods from one place to another; to find the most profitable ways, sites and conduits to do so, and to exploit the most remunerative markets. Moreover, these considerations equally led them to weigh the access to the techno-political complex that allows the movement of goods: the port, the road and the border, thus the basic components of the commercial corridor. It was this kind of commercial logic that enabled those opportunities that unfolded with the collapse of the state. This commercial logic – of movement, access and connectivity (in the sense of Scheele 2012) – places the road as well as the whole transit economy infrastructure at the core of local strategies. At stake for the import-export traders was, in these specific circumstances, the very risk to continue (or not) their economic activity, and, representing the other side of the token, the possibility to enormously gain from the new unfolding scenario. Negotiating access to the port, the road and to the border has always been crucial, but one of the very specific opportunities of Somaliland in the early 1990s was to establish the modes and forms of this access directly, without third parties or mediations.

16 Interviews with important local traders invariably started with preliminary remarks on the regional scenario, including the neighbouring countries in the Horn and in the Arabian peninsula. Strategic considerations normally involved European countries and also increasingly the Asian scene, including China, Thailand, Indonesia and Malaysia. The grand narratives in Wajale that I will describe in the last section of my chapter clearly mirror this form of presentation utilised by the important traders.

When recalling this period, the first ship (and the first trader) that arrived in Berbera after its liberation by the SNM troops in 1991 is often mentioned in Hargeysa. As one informant stated: 'Ibraahim Dheere [Ibraahim Cabdi Kaahin] sent the first ship to Berbera. He did it at his own expenses and without any insurance. He proved to all the other businessmen that it was still possible to use the port.' This act directly targeted the port of Berbera and the whole corridor and triggered the readjustment strategy of a number of important traders who were to become the new dominant business class in Somaliland. Agents brokered a deal with the local militia over the use of the port and re-activated a network of re-distribution of imported goods within Somaliland.[17] The specific interest to ensure the circulation of goods brought the road into focus. The obvious hurdles in this respect were the roadblocks, which entailed payments to each militia group along the way. At this point, commercial interests intersected with political ones. Before addressing this, however, let us further analyse the regional reconfigurations from the different knots along the trading route. For this, we need to move on to another post along the commercial corridors.

A Road's Detour: From Berbera to Djibouti and Back Again

Who were the important traders who profited from these changes and who actively operated to gain control of these processes? In examining these questions a detour via the traders' business trajectories and background is revealing – which also implies a detour towards another road. This will help us to understand how their careers have always been deeply connected to the corridors and road network of the region. Furthermore, it will also show that the dynamics of regional corridors should not only be assessed in terms of their linear connection between the sea and inland, but also in terms of a set of intersections and diversions. In particular, the detour draws our attention to another road, the connection Djibouti-Ethiopia (C3 in Map 5.1). I will show that, in this case too, it is the road and the political opportunities that make the businessmen.

The new scenario in the Somaliland of the 1990s displaced a number of well-known traders from northern Somalia who were linked to the old central state and its capital Mogadishu. In turn, other traders could better exploit the new challenges from less visible locations. These were in particular a group of Djibouti-based businessmen belonging to the Isaaq, the dominant

17 Another way to do this was by resuming the livestock exports from the country, as imported food was used to purchase livestock.

clan group in north-western Somalia. They had direct links with Somaliland, as they were connected with the groups in control of Berbera (on the basis of past commercial relationships and clan membership) and, in some cases, also with the SNM, having supported the rebellion movement during the 1980s. Besides these links, they could rely on credit given by Djibouti banks and they had been spared from the destruction and loss of resources that the war had brought about in Somalia. The deteriorating situation in Djibouti was an element that further persuaded them to consider other options. They had all previously migrated from Somalia to Djibouti to exploit different sets of opportunities. In addition to those that emerged from the French military base established in the country after Wold War II, many of the traders started to do business along the Djibouti corridor (mainly buying and selling goods) and in particular in the town of Dire Dawa located in the upper part of the corridor. After the rise to power of Siyaad Barre in 1969 and his socialist twist to the Somalian economy, an increasing number of Somaliland businessmen entered Djibouti and made their fortunes from the corridor. They chose Djibouti as the centre of their business and began to develop significant import companies. During the 1980s, Djibouti became a financial centre for these traders as well as for other Somali businessmen who benefitted from its conductive financial structure and its incentives for the transferal of funds (Marchal 1996) while trying to escape the strict state controls within Somalia. By the late 1980s, Djibouti had thus become part of a wider Somali economic space composed of actors operating beyond Somali national borders and using a range of opportunities in the formal and informal circuits.[18]

Above-mentioned Ibraahim Dheere, who sent the first ship to Berbera in 1991, belonged to this Djibouti group. His career started in Dire Dawa, 'trading vegetables to Djibouti like many petty traders', as his competitors readily recall, adding though, that 'already in the 1970s he was very big'. His move back to Berbera in 1991 incited more importers based in Djibouti to follow suit. During an interview with one of these Djibouti-based importers, it was remarked that

> In Berbera it was possible to unload a ship of rice in less than one week. Unthinkable in more crowded ports. But it was not only that. Djibouti has

18 In this sense, the collapse of the state in Somalia merely liberated groups and strategies that were already relying on parallel 'off-shore' circuits. The renewed post-war centrality of the Berbera corridor was only one of these dynamics. Other dynamics pertained for instance to the rise of remittance companies and their role in pushing forward the Somali technological revolution (Little 2003), the migratory movements and the establishment of long-distance international trading networks.

> no market, it is too little, in the past it was just an entry point for import and for the banks. After the war, in Somaliland it was all open (...) an open way to south Somalia and Ethiopia.

After Berbera (the port and the road) and Djibouti (the diversion), we now turn to the border zone between Somaliland and Ethiopia, which represents the final key element of the techno-political complex of mobility and connection. We will learn how, in the early 1990s, the Berbera route became the new axis of Somaliland.

The Inland Part of the Berbera Corridor: The Somaliland-Ethiopia Border Zone

Following the fall of the Mengistu regime in 1991, the new Ethiopian elite instigated a benevolent attitude towards the peripheral regions. Along the Somaliland-Ethiopia border, this approach was further reinforced through the presence of a number of refugee camps, which were the result of the Somali civil war and the related humanitarian assistance industry. The concentration of a large number of people in the camps and the assistance provided to refugees brought about many unexpected changes in as well as beyond the frontier region (Ciabarri 2008, 2011). Over the years, these changes gave birth to a new economy that was linked to humanitarian aid and based on the reselling of food rations in exchange for products not available in the camps. The camps became market centres where a number of basic products could be sold. Yet more importantly, spurred by the de facto re-opening of the frontier, the camps induced a revival of the trans-border trade.

The Hartashelkh camp in particular – located near the Berbera corridor – became the largest livestock market in the region in the early 1990s. Imported basic food items were exchanged for livestock, and on the basis of this activity, other import trades began to flourish as well. The big import-export traders described above were involved in these trades; but also medium and small-scale businessmen could resume their activity or start a business career. Whereas the trade began with items such as used clothes to be resold in the camps, later on business could include more sophisticated goods such as electronics coming from Dubai and destined for the Ethiopian market. The result of these transformations was an economic system based on appropriation and reselling of relief aid on the one hand, and unofficial trans-border trade on the other. During the course of the 1990s, the refugee camps turned from shelters against violence into centres of economic recovery for the surrounding region as Somaliland slowly returned to security and stability.

Hartasheikh, by virtue of its location on the road, operated as a kind of umbrella that covered the surrounding cross border trade. Positioned just off the main Ethiopia-Hargeysa-Berbera road (thus constituting another diversion) it became a crossroads of different pathways. From Hartasheikh departed the routes of the livestock trade from Ethiopia to Berbera and the imported goods from Berbera to Ethiopia. A regional circulation of food aid originated from here and a trade in used clothes, shoes and other basic products converged here. The money remitted by the diaspora abroad converged and departed and it was here that the communication systems that kept the refugees connected with this diaspora were to be found, including radios, landline telephones and faxes during the early days and, later on, mobile phones and internet.

When examining all these single knots along the route and their respective diversions – Berbera, Djibouti, Hartasheikh – what remains relevant from a road perspective is that, in order to fully understand the regional reconfigurations of social spaces and the related forms of mobility, circulation and connectivity, we cannot consider them separately, within a static perspective. It is necessary to combine them – just as the businessmen involved in these dynamics do, moving along the road physically and intellectually while imagining its prospects and producing effects.

The Political Encapsulation of the Infrastructure of Mobility: Producing a Free Space

The general reconfiguration of commercial relations in the region intersected with the political trajectory of Somaliland, particularly in terms of access to, and control of, the three major elements of the corridor: the road, the port and the border. Together, these three assets came to represent not only a commercial opportunity, promptly exploited by a few Somaliland businessmen, but also a political project. The success of President Maxamed Xaaji Ibraahim Cigaal in the Somaliland peace process of the early 1990s was founded on his ability to bring the revenues of the port of Berbera under the umbrella of the state. Furthermore, he was able to design a viable political project for Somaliland with the commercial corridor serving as its backbone.[19] Part and parcel

19 Apart from the invaluable role played by the elders and other community and religious leaders in effecting internal peace both at local and national level and apart from the new Somaliland nationalism linked to the violent memories of the civil war, the success of

of this project were the business interests involved. The failure to serve these interests was the decisive factor that had brought about the downfall of the previous Tuur government.

Cabdiraxman Tuur was defeated in the 'sheep war', in which the stakes were the revenues from the Berbera port. The subsequent peace process led to the national clan-conference of Boraama in 1993, during which Cigaal was elected president of Somaliland, and it was then that the institutional framework of Somaliland emerged and that the peace-process took off eventually. Cigaal's success resided on a commercial pact with the group of prominent businessmen in Djibouti described above, who belonged to his larger tribal group (Isaaq/Habar Awal), and in particular with the man who sent the first ship to Berbera in 1991, Ibrahim Dheere. These businessmen financed the demobilisation of militias through the distribution of food rations, which followed in exchange for tax exemption. The militiamen who manned the roadblocks were absorbed into the national army and the same roadblocks became part of the state efforts to control the territory. The traders also provided loans to the administration (as well as money for Cigaal's political battles and consensus-building) and support for the printing of the new national currency, the Somaliland Shilling introduced in 1994 (Bradbury 2008: 112). For the businessmen, the advantages came in purely commercial terms, giving them the chance to operate in a pacified and free environment and, from the port and through the road, to gain access to the border and the Ethiopian market.

While this commercial pact is well known, there are different views concerning the role played by the different actors; the order of the events, and the relevance of the pact for Somaliland. As we have seen, for the traders, it was initially all a matter of business. Other local actors from Berbera stressed the role of local groups. The ex-mayor of the city mentioned above reported: 'we were in control of the port and we called Cigaal. He did not come, he was not interested. Only before the Boorama conference he accepted.' From their viewpoint, Cigaal just drew a political project from a situation that had already been created. He had a port, a road and an open border, as well as a militia that presided over it – he just put the umbrella of the state over it. But this

this political project should also be considered in evaluating the peace process. Cigaal's political skills in dealing with political opponents and building consensus as well as in managing the so called 'Airport war' of 1994–95 are other crucial elements. My intent here is not to downplay the role of elders, of SNM and nationalism, or the role of Ciigal, but to draw attention to the commercial factor, thus far not considered, and to its primary role in the formation of Somaliland and its very functioning.

was also consistent with Cigaal's political trajectory since the 1960s (and with his personal background, as son of an important businessman of the 1950s), which aimed at appeasing the relationships with Ethiopia while avoiding any conflict over trans-border commercial integration. The latter was quite a minority stance within the Somali nationalistic tradition at that time. It also speaks for his political skills that he started to act only when there were reasonable chances to succeed. Considering the support he had received from the businessmen, Cigaal's assertion in a BBC interview that Somaliland had the chances to become the first 'tiger of Africa' (replicating the Asian economic tigers) is not pure folklore. Pompous as it appears, it expresses the idea of a state whose main constituents are a port, a road, and cross-border relationships. State building has thus coincided with road regulation. Far from being only an administrative act, these regulations powerfully resonate with local historical memories and territorial specialisation. However, unevenly shared and thus extremely ambiguous, these memories could be drawn upon in Somaliland to formulate a sense of difference with respect to the other Somali territories, and furthermore to substantiate its own autonomy, thus combining state building with nation building.

Finally, the support given to the state by the Djibouti group became a fundamental element of the functioning of both the market and the state in Somaliland. Direct negotiations and a regime of exemption characterised the relationship between the state and big businessmen. Furthermore, the pact resulted in a specific appropriation of port incomes, involving the local administration and the presidency through a non-transparent budget. The pact thus laid the foundations for two kinds of monopolistic positions: on the one side, that of the businessmen with respect to the internal market and, on the other side, that of the Presidency with respect to the other constitutional powers. Ahmed Serwai's words about this new elite were thus not merely dictated by his exclusion from the system – quite the contrary, they stressed this as a central element of Somaliland. As he noted, the competition among these traders directly targeted the road and transport, and the purchase of a fleet of brand new Mercedes trucks just revealed the attempt by one of the currently most successful traders to gain full control of mobility along the road.

Limits of the Road, Limits to Development

The sense of a free-wheeling and unlimited economy – substantiated by the accomplished free movement of goods and people and by free border crossing along the corridor – dominated the region throughout the 1990s. It nourished hopes for development and traders remember these times with a mixture of amazement and regret, hinting at unrepeatable profits.

By the end of the 1990s, the activities of big traders combined with the positive effects of other levels of trade in post-war Somaliland, which included small and medium businessmen in international trading networks for importing electronics, furniture and other goods for the Ethiopian and internal market. It also combined with the effects of a growing remittances economy (after the 1970s–1980s wave described by Jamal 1988) and the general reconstruction of the country. Trade, reconstruction and remittances developed to the extent of creating a real boom, which has had and continues to have profound effects on the Somaliland social landscape (Ahmed 2000; Bradbury 2008; Ciabarri 2010).

Other signs began to appear as well though. In particular, the unrestricted movement of people and goods along the road raised new concerns on the Ethiopian side regarding the control of the cross-border economy and security issues. This inaugurated another phase in the biography of the road marked – at least in the frontier zone – by a general atmosphere of precariousness and suspicion of people and goods in transit. Whereas the 1990s featured a sense of openness and freedom, in the subsequent decade the road became also a synonym for closure, contagion and danger.

The first sign of this change began with the clamping down of the Hartasheikh market. As described above, in the mid-1990s Hartasheikh had become a huge market place following the transformations brought about by the refugee assistance economy and the renewed trans-border trade. During these years, the system had enjoyed the benevolent attitude of the new Ethiopian regime, which permitted, and probably also benefitted from the booming cross-border trade. Their attitude, however, changed in the course of the 2000s. Though beneficial to both sides, the system also meant a huge loss of income for Ethiopia in the form of custom fees due to a flourishing smuggling centre that flooded its national territory with illegal items, in turn, giving rise to non-transparent money transactions. Given the restored peaceful conditions in Somaliland, in the late 1990s and early 2000s UNHCR carried out a series of repatriation programs in order to close the camp and in 2002, the last official refugees left the camp along with the humanitarian industry. It marked the return to normality in the Ethiopian border zone and subsequently revived the centralistic tradition of the government while paving the way to dismantling the market place. Again, what is interesting from a road perspective is the peculiar link established by the Ethiopian authorities, namely that the same road that symbolised growth and development on the Somali side, led to risk and danger on the Ethiopian side.

The Ethiopian authorities sent the army into the border zone in support of the custom officers. In Hartasheikh and Jijiga, goods, trucks and commercial vehicles were confiscated and several shops were closed. Economic activity was reportedly curtailed by 30% to 50% (EC Kenya 2003). Estimations from the side of the Somali traders do not give any precise amount but generally agree that this marked the end of an era. The subsequent withdrawal of the army however, left the custom officers once again without sufficient resources to prevent illegal trade. The diversion of Hartasheikh was directed back to the normal path. Yet, this generated new diversions.

The trans-border trade did not stop, but moved in part to the town of Wajale, which, in a way, meant it was returning to its classical location, as the history of the town has demonstrated. Wajale is located at the intersection of the Berbera road and the border, straddling both sides of the border of Somaliland and Ethiopia. In 2008 it had all the appearances of a new town, with an unplanned and confused set of streets and shops. It was dusty in the dry season and muddy in the rainy season and it lacked most of the basic services. In effect, the early 2000s marked a new beginning within a history made up of destructions and rebirths, as its inhabitants were proud to illustrate. The town itself is literally located in the middle of nowhere. Except for an elementary school, a customs house and a police station, it had virtually no services other than wholesale and retail stores, along with some hotels, garages and the gigantic mobile phone towers of Somali telecommunication companies.

The general impression of neglect stood in stark contrast with the colourful walls of the stores, which displayed huge advertisements for the goods on sale. As many of the stores were selling spare parts for lorries and cars, these wall paintings depicted trucks and the road. Oversized paintings of mobile phones, computers, TVs, refrigerators and other domestic appliances were also on view, along with furniture, cosmetics, bags of pasta and rice, cooking oil and cigarettes. The business community represented the largest community in the town, including sellers, agents and representatives of the trading companies, truck drivers, helpers and facilitators, young women catering for the housework and cooking, mechanics, hotel and restaurant clerks, and sex workers. Some policemen of the local police station and customs officers complemented this assorted community. However, travellers and commuters outnumbered the local residents by far. One of the busiest places in town was the immigration office, which recorded people entering or leaving Somaliland. High-speed cars transporting *kat*[20] across the border further completed the full picture, as

20 Also spelt *khat*. In Somali: *qaad* or *jaad*.

did what could potentially be found on the other side of the border, such as alcohol (Ethiopian beer and gin in particular), more sex workers, minibuses to Jijiga, better trucks, formal banks issuing letters of credit and an asphalted road.

Road, border and trade all featured prominently in people's daily conversations. Even though business tended to be conducted either on a personal basis or by specific business groups and alliances, the town has curiously generated

FIGURE 5.1 *Wajale*
SOURCE: L. CIABARRI, 2008

FIGURE 5.2 *Wajale: wall paintings depicting trucks*
SOURCE: L. CIABARRI, 2008

a number of 'grand narratives' of Wajale. These were collectively shared and recurrently repeated by its inhabitants and guests during the long afternoon *kat* sessions. The typical nature of these narratives constitutes a sort of collective representation of the town as they provide an account of its cyclical fall and rebirth, which invariably follows one of the following plots, as it was reported to me in a collective interview (April 2008):

> Wajale was destroyed in 1964 during the short war between Ethiopia and Somalia, which followed months of mounting tension along the frontier. In 1977–78 it was again directly involved in the Ogaden war, as its surroundings were one of the main theatres of the conflict. In the 1980s, the town – like any border area in the north – became progressively involved in the struggle between the Somali central government and the SNM, until it finally found itself at the centre of the civil war which erupted in 1988.

Or:

> In peaceful times the town has always recovered its position as a business centre for the trans-border trade. It has always been like this, since the village was founded at the time when the border between Ethiopia and the British Protectorate was definitely demarcated in 1954.

When compared with Hartasheikh, Wajale was diminishing. It is still evident though, that as of today, it remains a major commercial centre for Somaliland. It receives containers from Dubai, the Far East and China through the Berbera road, and then re-distributes these goods within Somaliland and across the border, thus representing a central knot in the local infrastructure of mobility and connection.[21]

Other grand narratives were about trade and the road. Talks about the road were however more strategical and just whispered. They referred to the possibility of reaching Ethiopia, to border controls and to relationships between the groups living on the two sides of the border; the latter being a factor that could both facilitate border crossing as well as hinder it. The recurrent aspect of these talks did not reside in the content, which was continually changing, but in the permanent circulation of information, knowledge and news about any factor related to road use and border crossing.

---

21 Recent observations have confirmed the conclusions drawn in 2008.

Whereas the curbing of Hartasheikh left Somaliland traders with an increasing sense of insecurity and suspicion towards Ethiopian intentions, the persistence of contrabandism further nourished the Ethiopian sense of paranoia and nervousness. Commercial flows along the road – involving goods, money and traders – became shrouded in opacity. Did money and goods come from formal or informal deals? Were people crossing the border involved in contraband business or were they simple visitors or businessmen linked to the formal economy? Of course the two considerations are not easily distinguishable, and they do not even pertain to different people. It is precisely these ambiguities that contribute to the sense of opacity.

These developments have to be placed within a larger political context, including the activity of opposition movements inside Ethiopia on the one side, and the radicalisation of the political situation in southern Somalia in the 2000s on the other. A constant trait of Ethiopian attitudes towards the Somali situation after state collapse has been the preoccupation with the strengthening of radical Islamic groups and the influence these groups could have on opposition groups inside Ethiopia. Trans-border trade and black markets are suspected to be potential resources for these oppositional groups and the movement of people is seen in this light as an uncontrollable means for their infiltration into Ethiopia (Medhane 2002).

Within the border zone, the road has thus emerged as a conduit for contagion (for similar occurrences, see Nugent and Asiwaju 1996; Masquelier 2002). By 2008 the entire border region was militarised. The new atmosphere in this region was characterised by harsh controls of people and vehicles crossing the border; by the confiscation of trucks, as well as systematic stop and search tactics employed by Ethiopian police in Jijiga upon the Somali population. In the last phase of the biography of the road taken into account here, the road became one of the preferred and most visible targets where control and repression were carried out.

Conclusion

Roads are often described as vital arteries for a state or a region's economic and social life. However, associating the biography of the Berbera road – made up of the histories of the various places found along it, of the diversions and of their overall connectivity – to the biography of Somaliland points to a much more intimate connection. A focus on the roads and commercial corridors in the analysis of the political re-configurations that occurred in the northern Horn of Africa since the early 1990s highlights specific aspects of the local processes of state building and of the reshaping of political and economic spaces. Yet there is something more to it than the simple provision of a material infrastructure to a political or economic project. The intricate web of commercial corridors in the region corresponds to an equally intricate interlacing between these routes, local histories and the trajectories of individuals and groups.

Moreover, the patterns of economic and political behaviour analysed here present an incorporation of the set of opportunities and potentialities conveyed by the road. In considering the long-term relationships between a road and a territory, the sense of the intimate connection between the two becomes even more evident. Commercial corridors appear as powerful forms of spatial organisation. They represent specific forms of spatial arrangements – to borrow Mbembe's (2000: 274) use of the expression in his exploration of the emerging geographies of Africa in the 1990s beyond state organisation, beside other possible forms such as borders, border zones, peripheries, centres, regions, cities, nomadic spaces etc. As such, commercial corridors produce and reframe spaces, articulate geopolitical projects; are a scarce resource as well as an opportunity for many, and they stipulate a set of behaviours in politics and economic activities. An ethnographic analysis of the road effects means not only to 'zoom in' or to enrich the general description by more detail. Rather, it provides a qualitatively different representation. By moving along the commercial corridor, re-constructing its biography and connecting road stories to locally significant social dynamics and phenomena, what emerges is a history of identities and territories in which the road occupies a central place.

References

Abir, Mordechai. 1980. *Ethiopia and the Red Sea*. London: Routledge.

Ahmed, Ismail. 2000. 'Remittances and their Economic Impact in Post-War Somaliland.' *Disasters* 24 (4): 380–389.

Aronson, Dan R. 1980. 'Kinsmen and Comrades: Toward a Class Analysis of the Somali Pastoral Sector.' *Nomadic Peoples* 7: 14–24.

Bradbury, Mark. 2008. *Becoming Somaliland*. Bloomington: Indiana University Press.

Burton, Richard F. 1987 (1856). *First Footsteps in East Africa or, An Exploration of Harar*. New York: Dover Publications Inc.

Ciabarri, Luca. 2008. 'Productivity of Refugee Camps: Social and Political Dynamics from the Somaliland-Ethiopia Border (1988–2001).' *Afrika Spectrum* 43 (1): 67–90.

Ciabarri, Luca. 2010. 'Trade, Lineages, Inequalities: Twists in the Northern Somali Path to Modernity.' In *Milk and Peace, Drought and War: Somali Culture, Society, and Politics*, edited by Markus V. Hoehne and Virginia Luling, 67–86. London: Hurst.

Ciabarri, Luca. 2011. 'Hargeysa: La ricostruzione della nuova capitale del Somaliland tra diaspora e campi rifugiati.' In *Antropologi in città*, edited by S. Allovio. Milan: Edizioni Unicopli.

Dalaklogou, Dimitris. 2010. 'The Road: An Ethnography of the Albanian-Greek Cross-Border Motorway.' *American Ethnologist* 37 (1): 132–149.

[EC Kenya] Delegation of the European Commission in Kenya. 2003. *Pre-Feasibility Study of the Regional Transport Sector in the Berbera Corridor.* Report. Nairobi.

Farah, Ahmed Y. and Ioan M. Lewis. 1997. 'Making Peace in Somaliland.' *Cahiers d'Etudes Africaines* 37 (2): 349–377.

Freitag, Ulrike. 2003. *Indian Ocean Migrants and State Formation in Hadhramaut.* Leiden: Brill.

Geshekter, Charles L. 1985. 'Anti-Colonialism and Class Formation: The Horn of Africa before 1950.' *The International Journal of African Historical Studies* 18 (1): 1–32.

Gewald, Jan-Bart, Sabine Luning and Klaas van Walraven (eds). 2009. *The Speed of Change: Motor Vehicles and People in Africa, 1890–2000.* Leiden: Brill.

Harvey, Penny. 2005. 'The Materiality of State-Effects: An Ethnography of a Road in the Peruvian Andes.' In *State Formation: Anthropological Explorations*, edited by Christian Krohn-Hansen and Knut Nustad, 123–141. London: Pluto Press.

Hersi, Ali A. 1977. *The Arab Factor in Somali History: The Origins and the Development of Arab Enterprise and Cultural Influences in the Somali Peninsula.* PhD thesis, University of California.

Howard, Allen M. and Richard M. Shain (eds). 2005. *The Spatial Factor in African History: The Relationship of the Social, Material, and Perceptual.* Leiden: Brill.

Jamal, Vali. 1988. 'Somalia: Understanding an Unconventional Economy.' *Development & Change* 19 (2): 203–265.

Kopytoff, Igor. 1986. 'The Cultural Biography of Things: Commoditization as Process.' In *The Social Life of Things: Commodities in Cultural Perspective*, edited by Arjun Appadurai, 64–91. Cambridge: Cambridge University Press.

Lewis, Ioan M. 2002. *A Modern History of the Somali.* Oxford: James Currey.

Little, Peter D. 2003. *Somalia: Economy without State.* Oxford: James Currey.

Marchal, Roland. 1996. *The Post Civil War Somali Business Class.* Nairobi: European Commission/Somalia Unit.

Masquelier, Adeline. 2002. 'Road Mythographies: Space, Mobility, and the Historical Imagination in Postcolonial Niger.' *American Ethnologist* 29 (4): 829–856.

Mbembe, Achille. 2000. 'At the Edge of the World: Boundaries, Territoriality, and Sovereignty in Africa.' *Public Culture* 12 (1): 259–284.

Medhane, Tadesse. 2002. *Al-Ittihad: Political Islam and Black Economy in Somalia.* Addis Ababa: Mega Printing Enterprise.

Miran, Jonathan. 2010. *Red Sea Citizens: Cosmopolitan Society and Cultural Change in Massawa.* Bloomington: Indiana University Press.

Mitchell, Timothy. 1999. 'Society, Economy, and the State Effect.' In *State/Culture: State-Formation after the Cultural Turn*, edited by George Steinmetz, 76–97. Ithaca: Cornell University Press.

Munn, Nancy. 1990. 'Constructing Regional Worlds in Experience: Kula Exchange, Witchcraft and Gawan Local Events.' *Man (New Series)* 25 (1): 1–17.

Nugent, Paul and Asiwaju, Anthony I. (1996), *African Boundaries: Barriers, Conduits and Opportunities*. London: Frances Pinter.

Pina-Cabral, Joao. 1987. 'Paved Roads and Enchanted Mooresses: The Perception of the Past Among the Peasant Population of the Alto-Minho.' *Man (New Series)* 22 (4): 715–735.

Robecchi Bricchetti, Luigi. 1896. *Nell'Harrar*. Milano: Galli.

Samatar, Abdi I. 1989. *The State and the Rural Transformations in Northern Somalia, 1884–1986*. Madison: University of Wisconsin Press.

Scheele, Judith. 2012. *Smugglers and Saints of the Sahara: Regional Connectivity in the Twentieth Century*. Cambridge: Cambridge University Press.

Schlee, Günther. 2003. 'Redrawing the Map of the Horn: the Politics of Difference.' *Africa* 37 (3): 343–368.

Swayne, Harald G. 1903. *Seventeen Trips through Somaliland and a Visit to Abyssinia: with Supplementary Preface on the 'Mad Mullah' Risings*. London: R. Ward Limited.

CHAPTER 6

Cosmological Work at the Crossroads

Commercial Motorbike Riders in Makeni, Sierra Leone

Michael Bürge

> But still the crossroads does have a certain dangerous potency; dangerous because a man might perish there wrestling with multiple-headed spirits, but also he might be lucky and return to his people with the boon of prophetic vision.
>
> CHINUA ACHEBE, *Morning Yet on Creation Day* (1975: 67)

•••

> Here at the junction we can see what is going on on the road. Nobody can leave this town without passing by. We see all the movements, we see who is coming in and who is leaving the town. We get our passengers here. Or we can get a job. [...] We have learned our lesson, our eyes are open, that's why the place is called *Opin Ay* ['Open Eye']. We have everything under control. Nobody can fool us again [...] We are seen, if we want that people see us. But if we want to hide, we can escape, just like this. [...] We know the roads, we can use them, but we can also ride our local paths.
>
> GIBRILLA, *Okadaman*[1], MAKENI, 2007

∴

Crossroads are central check-points in northern Sierra Leone's social environment. At crossroads, people check, that is, they observe and inspect whilst they are also being checked by others. At these thresholds of towns and villages they observe other people and goods travelling on the roads, entering and leaving the community and coalescing at the crossroads. Here people scout out

1 *Okadaman den* (singular *Okadaman*) is the term for commercial motorbike (*okada*) riders in Sierra Leone's *lingua franca*, Krio. In the following I will use the English plural form *Okadamen*. For the history of *okada* see Bürge (2009, 2011), Bürge and Peters (2010) and Menzel (2011). For the sake of anonymity of my interlocutors, all names in this chapter have been changed.

 | DOI 10.1163/9789004339040_007

for jobs or business. Others find their audience at the crossroads in order to preach about religious and moral duties. At the crossroads, people interrogate their positions and those of others in the world, assess the impacts of current developments and discuss the possible paths to take in order to participate in these developments and to improve their lives. Yet it is also at crossroads that people feel that they are interrogated and endangered by various threatening forces that hamper their (social) mobility. These forces critically interfere with people's trajectories, possibly leading them into traps of immobility. In this chapter I argue that crossroads are condensed spaces of ambivalence and indeterminacy, that is, of possibilities *and* impossibilities that one has to discern. I show how the ambivalence of the crossroads in Makeni, the capital of Sierra Leone's Northern Province, is closely related to the indeterminate position of the most prominent actors found at the crossroads: *Okadamen*, young men who ride motorbike-taxis in order to earn a living.[2]

Michael Jackson (1990) argues that in northern Sierra Leone, human and non-human actors – tools, machines, spirits, localities and also crossroads – *performatively* realise their 'personhood' in mutually constituting social practices and relationships (see also Ingold 2011). 'The crossroads', Aiyejina (2009: 9) writes 'is the meeting place of all realms – physical and metaphysical.' They are made up by the changing powers of various actors and practices that conflate at them. At the same time, crossroads are constitutive of these actors and practices. The relationship is dialectical. It is in this sense that crossroads and *Okadamen* form inextricably related actors.

Crossroads and roads in general are of particular significance in northern Sierra Leone. Crossroads, for instance, are the points of departure of various kinds of travels. Taxis leave for Freetown where relatives live with good jobs and money or where people have to go for bureaucratic issues. Crossroads are also points of arrival. Agricultural products arrive on lorries together with people from the villages. Indeed, Makeni itself is a big crossroad. It is the biggest town in northern Sierra Leone, where two of the most important highways pass through. Crossroads are passage points where roads shoot through, carrying along and circulating goods and people. Yet vehicles, people and goods do not only rush through and on (cross)roads. They also interact with and leave their traces on roads and crossroads and so shape these emerging spaces (see Ingold 2011).

'Money is in the streets' people told me, referring to the important role that roads play in finding a job and earning a living. There are those who walk or run along the roads for selling goods (see Klaeger 2012). Others in their shacks or bars try to benefit from the hustle and bustle of roads and crossroads

2 Fieldwork in northern Sierra Leone was conducted from May to November 2007, October 2011 to May 2012, and November 2012 to June 2013.

in particular. People also patrol roads for information about jobs or housing, for food and other daily needs. Crossroads are points of reference and 'fix-points' where many different activities, services, people and goods condense. Yet roads may also have a destructive impact on lives. People are involved in accidents or even die on the road, while others die in ambulances on their way to the hospital. People lose their houses and fields due to road constructions, while transport and other road-related business may turn out to be fallacious. Makeni's residents know about this ambiguity of roads and about the need to 'fight and suffer' in the streets for survival. Roads and crossroads thus form particular public spaces in which socio-economic and political struggles take place.

Crossroads are not simply intersections of strips of tar, just as tarred roads are not simply neutral material interventions in a landscape. Roads generally 'elicit powerful temporal imaginaries, holding out the promise (or threat) of future connectivity, while also articulating the political and material histories that often render these otherwise mundane spaces so controversial' (Dalakoglou and Harvey 2012: 460). Roads – and even more so crossroads – are socially and culturally *signified* and *signifying* spaces, inscribed with multiple meanings and histories. Crossroads and the multiple forces that conflate at them play an important role in the cosmological work across West Africa. Euba (1996) and Achebe (1975), for instance, depict the ambivalent powers of crossroads as they bring into prominence cosmological and historical features (see also Green-Simms 2010; Bakhtin 1981). In a similar vein, Bakare-Yusuf and Weate (2005), Aiyejina (2009) and Stoller (2002) explore the unsettling powers of crossroads as the ultimate meeting point of tricksters, spirits, deities and other transmogrifying figures who stand for the potentiality of both prosperity and destruction.

Crossroads are central elements in Sierra Leoneans' *cosmological work*. By cosmological work I refer to 'the work of constructing and reconstructing the (not always material) everyday worlds [people] inhabit' (Bastian 1998: 111). Sierra Leoneans make sense of their social environment *at* and *around* the crossroads. Here, the necessity, possibility and meaningfulness of 'checking' their trajectories increase. This is central to the cosmology of Makeni's residents within which crossroads emerge not only as the locations, but also as the object and subject of meaningful practices.

Okadamen are the other central actors in the interrelated processes of production of physical and social spaces and of personhood. In this chapter, *Okadamen* appear as a coherent and uniform group of young men. They may indeed be defined by their occupational role – as men who are paid for transporting goods and people on a motorbike. Yet there is great diversity, differentiation and stratification among the riders (Bürge 2011, 2009; Bürge and Peters

2010). *Okadamen* in Makeni come from different social, ethnic and economic backgrounds. Some used to be farmers or diamond diggers, others are schoolboys earning money in order to continue their studies. Some bike riders own the motorbike they ride, others even own two or three motorbikes, which they rent to other young men. Motorbike riders are organised in the *Bike Riders Union* and in the different parking grounds where they wait for passengers. Yet the majority of *Okadamen* rarely ever identify themselves as such. Those who do, emphasize the contingent and transient character of this occupation. They refer to other positions within the community and to other aspirations in their lives. They are an indeterminate group of people. Nevertheless, writing about *Okadamen* as a distinct and uniform group is not my own arbitrary decision. People in Makeni selectively reduce young men riding motorbikes to a single facet of their occupational identity. This facet is mainly – but not exclusively – tied to a negative interpretation of their role within the community. In the following, I explore how *Okadamen* and crossroads mutually constitute each other's *ambivalent forces*.

The analysis I develop in this chapter draws on observations made at various junctions across town. I focus mainly on one particular crossroads in Makeni, known as *Opin Ay*, 'where the eyes are open'. *Opin Ay* links the highway connecting Freetown with the diamondiferous regions in eastern Kono District and the road leading into the centre of Makeni, the most important business centre in the north. I argue that the practices and discussions carried out at this junction stand to a certain extent *metonymically* to the general – social, political, economic and moral – situation in Makeni and in Sierra Leone. Furthermore I show that according to the local cosmology, crossroads are places where powerful actors – diviners, sorcerers, hunters and tricksters – congregate and try to tame powers of the visible and invisible worlds, to incorporate and make them productive for the community – or pretend to do so. By embodying the ambivalent powers of (cross)roads, these actors themselves become figures of indeterminacy. Depending on the configuration of powers, their relational 'personality' changes. Crossroads are the realm of shape-shifting figures who can be seen as heroic saviours of the community sacrificing themselves for its survival (see Ferme 2001). Shape-shifting figures can also serve as critical 'catalyst[s] for dialectic self-examination' and against self-complacency to push the community forward for continuous improvement (see Aiyejina 2009: 7); and they can dissimulate their real identity, betraying and misguiding the community. Shape-shifting trickster figures appoint themselves as saviours, yet also turn innocent people into scapegoats and sacrifice the community for their own benefits. Makeni's commercial motorbike riders, the *Okadamen*, can be seen as embodying and realising these indeterminate powers in the present.

Congregating at crossroads and roaming northern Sierra Leone's roads, *Okadamen* evoke and embody ambivalent notions of shape-shifting trickster figures – staying in-between, bridging and provoking fragmentation and fissures in the community (Bürge 2011). In Makeni's contemporary atmosphere of promise and uncertainty, *Okadamen* consider themselves as the community's saviours in that they they suffer on the road and sacrifice their bodies to provide mobility and 'development' to the community. Not everybody in Sierra Leone agrees on this representation, but see *Okadamen* as dangerous and deceiving actors who dissimulate their 'real identity' and cause confusion. Some consider *Okadamen* to be responsible for various social malaises often linked to (hyper)mobility (see Kanu 2012). *Okadamen* and crossroads are quintessentially indeterminate, they can be seen as both good and bad and as creating both order and disorder. Under current difficult social conditions, however, Makeni's residents tend to emphasise the negative impacts of *Okadamen*. As I will show below, this ascription is not arbitrary. It relates, rather, to the resemblance and continuity of *Okadamen* with historical and cosmological forces that roam about roads and that try to master crossroads, promising security yet also threatening destruction.

Anthropological Interests in Roads and Crossroads

In recent years, anthropologists have paid increased attention to the multifaceted ambivalences of roads and road-related activities and discourses in Africa (Klaeger 2013: 360). In her seminal article on 'road mythographies' surrounding Niger's Route Nationale 1, Adeline Masquelier, for instance, paints a picture of the 'profoundly contradictory nature of roads', of roads as 'a space of both *fear* and *desire*' (2002: 831, emphasis added). More generally, the recent volumes *Roads and Anthropology* (Dalakoglou and Harvey, 2012) and *Ethnographies of the Road* (Klaeger 2013) emphasise the fruitfulness of roads as lenses for ethnographic investigations into current socio-cultural conditions. African roads in particular have become objects for investigating people's everyday struggles with infrastructural improvement, technological threats, speed and (often related) deaths (Lamont 2013; Sanders 2008).

Looking into roads in Sierra Leone involves dealing with numerous ambivalent issues as well. For instance, 'substandard' roads cause slow and exhausting travel as well as traffic congestion in Freetown and are often criticised as obstacles to development. At the same time, exaggerated speed enabled by tarred roads regularly provokes outcries due to the numerous speed-related fatalities (which may also be linked to drivers' haste, their drug consumption, insufficient vehicle safety, and so forth). In the country's north, ambivalent discussions of mobility have deep historical roots. As Rosalind Shaw (2002: 93) explains, 'roads acquired a powerfully ambivalent potential in Temneland as channels of both trade and slave-raiding, wealth and disappearance'. These specific local histories live on in present imaginations and in performances *at* (cross)roads, and they became 'potent metaphors for an "immoral economy" of colonial and post-colonial commerce and transport' (Shaw 2002: 64).

In the following section I trace how the ambivalent potentiality of roads and crossroads in northern Sierra Leone is historically constituted and presently evoked. I then outline how *Okadamen*, who visibly dominate the crossroads in Makeni, re-enact these ambivalent forces at the crossroads.

Interrupted and Departing Roads

In *Memories of the Slave Trade*, Shaw (2002) shows how past experiences of (violent) intrusions from the Atlantic and Islamic worlds reverberate in Temne ritual and historical memory. They inform the production and perception of today's places. The ambivalent potency of crossroads is historically constituted. Crossroads are the 'chronotopic materialisation' (Shaw 2002: 63) of an ambivalent past in today's landscape. This does not mean, however, that crossroads are simply the present materialisation of past *facts*. They are discursive and performative configurations for remembering the past and therefore form part of contested (memory) politics. Their meaning is realised in particular evocations of past features.

In the cosmology of Temne people in northern Sierra Leone, the ambivalent potentialities of prosperity and destruction of roads (*ta-soŋ*) are intensified where roads come together, cross (*pəŋkinɛ*) and are '[channelled] from multiple directions', according to Shaw (2002: 94).[3] *Pəŋkinɛ* derives from *pəŋk*, which denotes states of 'being crossed', that is, states of 'interrupted' and 'disturbed' flows. Before I became interested in crossroads, I knew of the term *pəŋk* in connection with mentally disturbed people. Someone's physical and mental state might be crossed, confused, stirred up and thus stand *in-between,* possessed by spirits – a risk lingering particularly at 'crossed roads' (*ta-soŋ tə-pəŋkinɛ*). *Pəŋk* stands for neither being in this present and visible world (*nɔ-ru*), nor fully in the invisible, dark and powerful world of spirits, witches and the dead that converge at crossroads.[4]

3 Jackson (1977, 1990) argues similarly for the Kuranko in northern Sierra Leone.

4 Littlejohn (1963) distinguishes between four worlds in Temne cosmology whilst my interlocutors distinguished in a more Manichaean manner between visible/invisible, living/dead, now/before, bush/village.

Saidu, a Ka-Temne teacher, explained to me that '*pəŋk* is where the road is interrupted by something going like this', making a sign with his hand from the left to the right and back, 'or somebody stands in your way, disturbing and blocking you.' He went on:

> '*Pəŋk*' is like a cross, a man who lies like this in his bed, not allowing his wife to lie down. This is wickedness. But Jesus was also '*pəŋk*', on the cross. Or you know, your head can be '*pəŋk*', crazy, it's always something strange. You have to be careful at these places, you know that they make sacrifices there, the place is very powerful and people there can disturb you.
>
> MATENE, 2012

This resonates with Shaw's (2002: 49, 93) findings that at crossroads 'people can come from different directions to harm you' and 'bush spirits and witches travelling along [the roads] at night, congregating at crossroads [...] sometimes "seize" (*wop*) people [...] making their victims crazy and disoriented (*pəŋk*)'. My interlocutors usually called crossroads *ta-soŋ sə-səkənɛ*, meaning 'the road departs', 'disperses' or 'separates', but also 'changes' or 'becomes'. Maberr, one of my riding friends, explained to me the three directions of the separating roads: 'When you move on a road, you come to a point where one [road] goes like this, one like this, one like this, *ta-soŋ sə-səkənɛ*.' He then continued:

> The road you come from is behind you. You cannot go back. You have to go ahead. At this point you have to decide which direction to take. It's difficult, because sometimes you don't know where to go. [...] Each direction is possible. [...] Like in life. Sometimes you don't know which is good and which is bad. Perhaps there are potholes, or they go where you don't want to go, somebody waits for you around the next corner. You can perhaps ask the people around. They can advise you. But you cannot be safe at this place, some of them also fool you. They confuse you and, you know, people in this country have *bad at*,[5] they will try to destroy you, advise you the wrong way. So you have to be careful. You have different choices, but you cannot know what happens to you.

---

5 *Bad at* ('bad heart') means jealousy, envy, leading to attempts to hamper others' development by pulling them down (*pul doŋ sindrom*) (Bolten 2008). The condemnation of *bad at* and *pul doŋ sindrom* as *the* ubiquitous social and personal ill hampering the country's development (*fɔ nor go bifo*) found its way into newspaper articles and popular culture, expressed in songs like *'Bad At'* by musician Emmerson Bockarie or *'Nor pwell me'* ('Don't destroy me') by Treasure.

> You have to go ahead, you cannot go back, you cannot let them *pul yu doŋ.*
>
> MASINGBI, 2011

It is a challenging place indeed. Yet at the crossroads one also finds opportunities for jobs or can meet people who might change one's life trajectory, whether for good or for bad. It can be "'*a trapping place*" or a central node in which you can find "*all the messages you are looking for*"' (Shaw 2002: 94, original emphasis). It is a check-point, where flows of people, goods and different powers condense. Mastering the crossroads, deciding over the direction of movements, interrupting and channelling the flows of others (instead of being interrupted or trapped) means control. Yet control is fleeting and sought after by different people and forces. Tamed powers can be used for different purposes and the presumed control of these powers might turn into being controlled by them. This cosmological ambivalence of the crossroads is informed by the indeterminate role that roads – and the powers they convey – played throughout the history of the region and its global integration.

In Sierra Leone, goods and people have moved up and down roads and rivers for centuries, and they brought both prosperity and despair. These movements intensified during the transatlantic slave trade and the 'legitimate trade' of the colonial era (see Wylie 1973). Those who controlled the roads and the transported goods benefited from the exchange. Those without control and unable to defend themselves were exploited or shipped as slaves (Shaw 2002: 89). At crossroads where roads, rivers and the sea come together, the deprivation of those exploited and the wealth of those controlling the commerce were particularly close.

After Sierra Leone's independence (1961), President Siaka Stevens (1971–1985) used the road network for selective political and economic integration of parts of the country. He cut off some regions and purposefully excluded their population and products from access to Freetown. Mariane Ferme (2001: 34–35) has shown how communities in the south rejected community labour for road and railway construction in order to defend their remoteness from intrusions and exploitation along these arteries during colonial and post-colonial times. During the civil war in the 1990s, roads were gateways for highly ambivalent and often destructive forces. Various factions moved quickly on the roads and entered villages and towns. They erected roadblocks where decisions were made regarding other people's movements, they harassed them or defended their communities. Generally speaking, roads – and their absence – played an ambivalent role in Sierra Leone and were subject to contestation and political struggles.

During my fieldwork in 2007 most people in Makeni wished for tarred roads, just as they longed for other modern technologies which would connect them to the rest of the country and to the world. Tarred roads were seen as a path toward a better life, whilst bad roads, missing railway connections and lack of electricity had become emblematic of the state's decade-long negligence of and disconnection with the town. People commented on this to underline how the post-war government of President Tejan Kabba (2002–2007) discriminated against the north for ethnic reasons and punished Makeni as it had served as the last stronghold of the *Revolutionary United Front* (RUF), the rebel forces of the civil war (1991–2002). People often complained about how late the highway between Freetown and Makeni had been tarred and about how it had only served to carry away goods and to send politicians and their thugs to Makeni. In their view, traffic and connectivity within the town and the Northern Province were still neglected by the government.

When I returned to Makeni in 2011, discourses had changed once again. Many residents were now praising the road constructions in Makeni, the 'perfect highway' between Freetown and Conakry and the railway built for transporting iron ore from Bumbuna, north of Makeni, to the seaport in Pepel. In their view, all of this had been realised by their new president, Ernest Bai Koroma, and was proof of his commitment to finally develop the north where he hailed from. At the same time, people of Makeni frequently reminded themselves of the price paid for these developments. For instance, whereas trains and lorries were carrying away tons of gravel, people questioned whether the gravel contained only iron ore, or if it also carried gold, diamonds and powerful spirits. Others complained that few people had been employed and that those that had were poorly paid, whilst the majority was loitering in the streets of mining towns, engaging in crime, violence and prostitution. People complained that mining companies did not pay taxes, as they bribed politicians with money, luxurious cars and houses. Furthermore, prices for rent and food had increased, which was perceived to be related to the influx of expatriates and new consumer goods. The costs for being connected to the wider world and for easier and quicker transport on newly tarred roads were financial pressures, social conflicts and environmental degradation. Blood was shed both on and for the new roads as they were capable of bringing social and physical death to Makeni. This reminded people of the ambivalence of roads and of their similarly ambivalent powers. Ultimately, it reminded them of the two-directional movements on the roads, the dichotomy between those who control and benefit from these movements and those who are excluded from its benefits, and thus from well-being.

Kevin, a car mechanic, driver and teacher in a vocational institute, explained to me that roads, cars and other 'foreign' inventions and knowledge were not genuinely European, but the result of Europeans being more skilled in appropriating – but also sharing – 'African' powers:

> We cannot produce cars, motorbikes and other machines. White men teach us now to build roads, but they plan them. We only use and repair these things, sometimes it's also just guess-work. [...] People in Africa can only buy and repair, but not invent. This is because the white men trick our spirits into bottles and bring them to Europe. They dance in the bottle, the spirits follow them and then they go out and lock the bottle. You put them into factories where they invent and produce for you all these machines.
>
> MAKENI, 2007

Shaw (2002: 53) refers to similar perceptions about uprooted spirits turning into menacing powers that had to be contained in the bush and at the margins of society. Europeans had been able to tame them and take them abroad where they invented and produced various machines, as Kevin and others explained to me. Therefore, today the spirits inhabit roads, vehicles and planes since they were part of the roads' production. Brought back to Africa, they threaten to spill back into villages and to attack people on the roads if they are not properly controlled.

These narratives seemed to be a means to criticise Europeans for exploiting Africa by high-jacking productive capacities while unleashing destructive powers. Many Sierra Leoneans, however, also blamed the jealousy and greed (*bad at*) of their fellow citizens for the current situation. Ordinary people were said to lack the power to control spirits, make inventions and to use them productively. And still, there were extra-ordinary locals with special abilities that enabled them to see hidden forces with their 'four eyes' (*ɛ-fɔr t-anlɛ*) or 'crossed eyes' (*ɛ-fɔr ɛ-pəŋkinɛ*). Diviners and hunters had the power to disguise and protect themselves and to enter their worlds at the crossroads of various worlds (see Shaw 2002: 93). Less powerful Sierra Leoneans entrusted their fate to these extra-ordinary figures that 'vaunt their powers and draw grudging admiration of those who know of them' (Jackson 1990: 59). The powerless hoped to participate in the success of the powerful in taming invisible forces. However, they often felt betrayed by the powerful who monopolised their power and used it for personal benefit. I was often told 'Our big men are greedy, they do not share their power like you people.' In their contemporary cosmological work, people included extra-ordinary figures at crossroads rather as betraying tricksters than as benefactors. *Okadamen*, the visibly dominating forces at the crossroads, were consequently placed in this kind of negative genealogy. From their own perspective, they saw themselves as self-sacrificing defenders of the community who were criticised by envious and destructive people with *bad at*.

Makeni's Crossroads: Open Eyes and Hidden Places

Sierra Leone's tarred roads are predominantly the domain of (young) men. They drive, walk, work and meet along the roads whilst women tend to work at home, at the market and use secondary tracks for transporting their goods from the villages to town. Traffic inside Makeni and at the busy crossroads around town is dominated by *Okadamen*. In the statement opening this chapter, Gibrilla explained the importance of the junction that people often call *Opin Ay*. At this entry point for Makeni, it was mainly young men whose 'eyes were open'. Here they controlled their own and other people's movements and thus the trajectories of their lives.[6] Waiting for passengers at that crucial crossroads, the young men checked people and things passing by. They were ready to interrupt their daily routine in an instant in order to do business, find a job, get their share of possible profit and, if needed, to deny or tax access to town (see Figures 6.2 and 6.3).

During my fieldwork carried out in 2007 many young men were waiting at the junction for the opportunity to ride a motorbike for a few hours or to rent one on a daily basis. Others engaged in community work, such as picking up litter or cutting down trees, hoping to be employed by NGOs or wealthy businessmen and politicians who paid money for various odd jobs and services. More senior riders were hired for transporting cannabis, gold, palm-oil or diamonds to Freetown or Guinea, and for smuggling new motorbikes and foodstuff from Guinea.

In 2011 another opportunity opened up at *Opin Ay*. The junction became the collecting point for the buses bringing workers to and from the various construction sites and plantations of multinational companies involved in mining and biofuel production around Makeni. At the crossroads one could meet foremen and other employees who could potentially open doors for employment.

6 Gibrilla gave me his assessment in 2007. In 2011, *Okadamen*'s presence at the junction had considerably changed and faded. This was clearly visible in the bikes' patterns of movement. During their working hours, *Okadamen* hardly ever stopped and waited for a passenger. They were constantly on the move, transporting and looking for the next task. Many of them actually worked for former *Okadamen* who now had opened their shops around *Opin Ay*.

FIGURE 6.1 Okadamen *at* Opin Ay
SOURCE: M. BÜRGE, 2013

The crossroads served as a point of departure and return. It was there that riders could satisfy their daily needs, including eating, refreshing themselves and getting a haircut as well as repairing and washing their motorbikes. At *Opin Ay* 'cinemas' showed football matches and videos, 'tele-centres' charged phones and music was often playing. Many riders had a nearby place where they could sleep and 'romanticise' women. They smoked, drank and gambled in the surrounding hidden places (*ghetto*).

In 2007 roughly 500 young men rode motorbikes in Makeni; in 2012 their union claimed that about 5,000 young men were earning their money on the town's roads. Some owned their bikes, others rented from wealthier businesspeople, NGO workers, police officers and government officials. In 2007 they paid a weekly rent (*mastamɔni*) of 120,000 Leones (40$), which was equivalent to 120 rides within the town. Riders started at about six o'clock in the morning and finished only when they had made enough money, usually after dusk. Some parked their bikes during the night; others tried to maximise profits by renting the bikes to other riders who worked during the night.[7] Yet the night

7 For more technical, organizational and financial aspects of commercial bike riding see Bürge (2009, 2011) and Menzel (2011).

shifts had become increasingly dangerous as *Okadamen* had occasionally been attacked and stabbed to death by armed robbers stealing motorbikes. In 2013 this led to a temporary suspension of transport activities during the night, which in turn incited *Okadamen* to stage protests against the 'useless police'.

Okadamen protected themselves and people in the area from unwanted intruders and tried to enmesh themselves into networks of wanted beneficiaries and benefactors. For this purpose, their eyes (and ears) had to be open in order to scrupulously check and decide whom to trust and whom to avoid. *Opin Ay* was the most important nodal point to do so in this part of town. Yet criticism and obstacles to their social mobility urged them to shift, enlarge and diversify their social relations and networks. Similar to situations of young men in other African locales, they tried to 'insinuate themselves into the lives of others [...] without becoming too obligated, too dependent or tied down' (Simone 2005: 519) to one particular place and by particular actors. They were organised in subgroups within the *Bike Riders Union* that should provide assistance in case of struggles with the police, medical treatment and wedding and funeral ceremonies. Bike riders participated in rotational credit and saving groups (*osusu*, pl. *ɛsusu*) which provided mutual assistance. Participating in an *osusu* allowed the riders to selectively knit social networks in which to participate as creditor and debtor. Thus it allowed choosing with whom to share money as well as whom to exclude from its benefits, at the junction or elsewhere (Bürge 2011: 79–81).

Okadamen did not wait for salvation from being unemployed and without money, a frustrating destiny of many young men in Sierra Leone and beyond (see Hansen 2005; Honwana 2012; Masquelier 2013). They sought betterment through movement. At the junction, *Okadamen* took breath and checked whether they needed to adjust their trajectories before they plunged onto the road again. Crossroads thus became an important point of passage rather than a point of departure or return for their everyday journeys within and beyond Makeni (see Figure 6.1).

Opin Ay was not only a point for observing the movements of others but also for being observed and 'checked'. In this sense, the junction could turn into a stage for performing one's commitment to the community's well-being. At the same time, a failed staging could also expose one's inactivity to the eyes of others. Having lived at the junction and spent much time with these young men, I could observe how people regularly criticised their way of living. On the one hand, young men who were sitting around 'idly' were often blamed for doing nothing and for being 'useless' – perhaps the worst insult for a Sierra Leonean, as Bolten also argues (2012a: 503). *Okadamen*, on the other hand, were frequently criticised for exaggerating and exceeding their activities, they were blamed for doing the wrong things. Bike riders were openly accused of

FIGURE 6.2 Okadamen *'checking' movements*
SOURCE: M. BÜRGE, 2013

over-speeding, over-loading and consuming drugs, which led to accidents.[8] They were also accused of penetrating every corner of the country and thus exposing themselves and the community to serious threats, contributing to the spread of sexually transmitted diseases and making 'impregnated' school girls drop out of school. Many of Makeni's residents were doubtful about the *Okadamen*'s commitment and sincerity and assumed that in the long run, they would emigrate and leave their close ones and their community behind.

Although *Okadamen* exposed themselves to the elements every day throughout the year and struggled to offer transport services on bad roads, their doing was not considered to be real 'work' with any sustainable impact on the community. People wanted *Okadamen* to go (back) to school, work on the fields or find a white-collar job in an office. Some condemned their sinful lifestyles and tried to convince them to respect their religious duties. Others made attempts to thwart *Okadamen* by denouncing them to the police or by pursuing

8 'Over-load' describes the *Okadamen*'s practice of carrying two pillions instead of one, or huge objects such as mattresses, furniture and animals. Among young people, it also had sexual connotations as *Okadamen* easily moved between their various partners' places and quickly transported girlfriends and prostitutes for other people.

them with sorcery and other 'African practices'. People linked virtually any social problem to *Okadamen*'s movements on their Indian-made motorbikes. '*Okadaising*' the society (Kanu 2012) became a widespread synonym for anti-social behaviour.

Such criticisms meant that it was not always appropriate for *Okadamen* to be seen. Only a few steps away from the open junction where everybody and everything was observable, there were back-alleys that led to hidden corners within and outside of town. In these secret places, *Okadamen* could take a break, socialise with people of their choice, drink *poyo* (palm-wine) or smoke *jamba* (marijuana). In times of jealousy and mistrust, solidarity among young men was often strengthened in places such as *Ro-mankne*, a 'hidden place' in local Temne, or *Ro-gbundu*, a 'secret place', where some of my bike-riding friends went to relax. Mamadu, who had taught me how to ride a motorbike, told me:

> Sometimes you need to know how to hide. You know, they are my brothers at the junction. But they have *bad mot* [bad mouth]. Sometimes you cannot see me, because I went to my village. Or I go to *Lukin Tɔŋ*,[9] where these people cannot see me. I can make *kul at* [cool heart] with my friends, with Ahmad or Conteh. Without being disturbed, but I can see everything and move.
>
> RO-MANKNE, 2007

Crossroads were physical and metaphysical bottlenecks and passage points that channelled ambiguous forces. Physical and metaphysical traffic congested the place. Moving forces blocked each other, they had to turn or just opt for a stop. The circulation of people and goods slowed down in this moment and place. At the same time, the density of people and activities increased. People otherwise overwhelmed by the increased speed of the world could grasp and approach here hypermobile forces such as the *Okadamen*. At the crossroads, they could find a large audience for effectively challenging and limiting other people's (social) mobility. They offered a public space for critically discussing and defiling other people's reputation, 'pulling them down' with envious gossip (*bad mot*). Practices and discourses of *bad mot* (gossip), *bad at* (envy) and the *pul doŋ sindrom* were rampant in Makeni.

Being challenged and provoked by others at the junction makes 'hearts warm'. Yet it is not only the heart that becomes 'warm', but also the entire body,

9 *Lukin Tɔŋ* ('watching the town', 'viewpoint') was a particular observation point and *ghetto* close to *Opin Ay* from which the junction was perfectly visible, though one could not be seen.

mind and nerves which are stirred up. *Wam at* might result in immediate violence or states of stagnant frustration and 'disgruntledness' (Bolten 2012a: 503). One has to find the right way to overcome this state of personal crisis in which one loses to a certain extent control of one's senses. Being guided by other people, consuming drugs and alcohol or avoiding provocative situations may help to restore a certain equanimity with the social environment (see Bürge 2011). As *Okadamen* disappeared and hid (in order to meet people whose presence did not 'warm the heart'), they continued to incite imaginations about their 'antisocial' and 'useless' activities. *Okadamen* were accused of wasting their potential by drinking *poyo*, smoking *jamba* or gambling. People even suspected them of harming society with harmful practices, such as by selling themselves and exposing the community to the dark and threatening forces of the 'underground', to criminals and witches. For the *Okadamen*, however, those same people who criticised and threatened their activities at the crossroads also became challenging intruders. Mutual distrust, envy and sabotage were often the outcome.

These destructive tactics and discursive figures of scapegoating have their historical antecedents. They are rooted in people's previous experience with mobile forces, (cross)roads and with related states of insecurity. The *Okadamen*'s position at crossroads was associated either with the rebels who controlled and harassed people at roadblocks and checkpoints in the past or with those who enabled economic enterprises to flourish after decades of government mismanagement. *Okadamen* brought pleasant luxury goods for example, just like the sixteenth century merchants, or they exploited people, similar to the slave traders of the same era. Judgement of the ambivalent powers was contingent, but not arbitrary.

Tricksters at the Crossroads

> We, the bike riders, make this place safe. […] Without us, Makeni is nothing, we are the economy of this place. You see the women at the market? Without bikes, they have to walk from their villages, walk under the hot sun. […] Without bikes this place is backward, more dark. So why people say we are dangerous? We give young people a possibility to earn good money, otherwise they would idle around, they would steal, because they are disgruntled; you know the history! But a bike rider is a serious person, he has responsibility, for his bike, for the community.
>
> SEMBU, *Okadaman*. MAKENI, 2007

FIGURE 6.3 Okadamen *waiting for passengers*
SOURCE: M. BÜRGE, 2007

Sembu was a member of the *Bike Riders Union Task Force*. This internal 'police force' had representatives at every junction and was responsible for 'disciplining' riders. Its mandate – although contested by many riders – was to ensure that all bikes and riders were registered with the *Bike Riders Union*, and that all riders carried a driver's license and number plate. The *Task Force* also controlled road safety and owersaw the proper conduct of the riders. In his statement Sembu combined parastatal security responsibilities with more 'informal' contributions of *Okadamen*, portraying bike riding as an activity that enhances safety and security within the community. He claimed that bike riding provided employment for young men, for those 'disgruntled' elements that had taken up arms in the conflict from 1991 to 2002, and thus prevented them from 'antisocial' behaviour – contrary to other people's accusations against *Okadamen*.[10]

Makeni used to be a 'dark' place, as people said. Until 2012, there was hardly any electricity, apart from small private generators. People also referred to

10 See Peters' work on youth, crisis and war in Sierra Leone (2011a, 2011b, 1998), as well as Hoffman (2011a, 2011b), Fanthorpe and Maconachie (2010) and Christensen and Utas (2008).

darkness in a more figurative sense: Makeni was backward and haunted by 'African' practices. People claimed that witchcraft directed against others was an outcome of jealousy, of people having *bad at* stemming from poverty and 'lack of development'. Additionally, witches and other dangerous forces needed the covering of physical darkness. As many bikes circulated during the night, some people saw this as proof of the fact that *Okadamen* worked with antisocial forces of the dark (*aŋ-sum*). In the statement above, Sembu argues against this however, and instead stresses that people were brought home safely by *Okadamen*.[11] They roamed the roads and illuminated them during the night, checked for intruders, reported threats of armed robbers to the police or settled the issues themselves. The police hired the *Okadamen* as 'watchdogs' as they went around Makeni hearing gossip and rumours. Bikes transported goods to the markets, accelerated the economy and spread information quickly.

With their motorbikes, *Okadamen* had introduced productive forces and enabled people in Makeni to participate in these flows and movements. They saw themselves as those who upheld social reproduction, developing and 'enlightening' Makeni comprehensively. Controlling the arterial roads as was the case at *Opin Ay*, they protected the community from the outside and from its dangers – the often disguised adverse forces which first needed to be identified. Before and after the election in 2007, anxieties regarding possible outbreaks of violence troubled the country. There were rumours of groups of 'strangers' entering villages and towns and harassing locals, as well as of people being kidnapped and used for rituals that were meant to enhance the powers of politicians. During election days *Okadamen* patrolled the town to intervene in cases of fraud; they moved between polling stations and observed the counting of votes to avoid manipulations by the SLPP, the party of the Southerners, the 'strangers'.

Okadamen risked their lives. They frequently fell on muddy tracks, were robbed and at times even killed during the night. But instead of being praised for these 'secular sacrifices' (Lamont 2013), their efforts to move and enable people to keep up with increased social speed were criticised as irresponsible. Instead of being protected by the police, such as by arresting armed robbers, police rather arrested *Okadamen* for missing number-plates. Upset about such injustices, in 2013 *Okadamen* violently protested on Makeni's roads and attacked a police station. Although they enlightened Makeni, *Okadamen* became

11 When commercial bike riding during the night was suspended due to regular attacks on *Okadamen*, people often desisted from moving during the night as they felt unsafe. Among expatriates not disposing of a car it was also a rule to only move by *okada* after dusk.

targets of 'dark African practices', that is witchcraft and witchcraft accusations. Daniel, a student and former *Okadaman*, explained to me:

> There was a time [...] when the riders at *Opin Ay* had many accidents. Suley fell down three times, Conteh twice. Bah broke his arm, others have scars on their faces. So the riders believed something wants to come to this junction. You know that we have spirits and witches which come along the roads. And they meet where the roads come together. That's why they made this sacrifice. They call it *Sathka*.[12] It's our culture. Some say it's Muslim. [...] You put rice, couscous, corn. You put everything you can eat. Then you give it to the people to eat. [...] You also put some on the road. [...] They put it where the road spreads, *ta-soŋ sə-səŋkinɛ*. [...] You know that in this country you have to protect yourself when you make business. People have *bad at*. [The sacrifices] make these powers, these fetishes, loose. [...] You ask a native doctor how best to do if you have problems.
>
> MAKENI, 2013

Okadamen were constantly provoked and degraded at the crossroads – and even bewitched. When they fell from their bikes, many of them did not ascribe this to 'accidents' or 'bad luck', but to purposefully being harmed by people with *bad at*. Remedy for these practices had to be found within the same logic of 'traditional' practices.

In the remainder of this chapter, I shed light on how today's ambiguous appraisal of activities around the crossroads are informed by past experiences inscribed in local cosmologies of the ambivalence of crossroads. The historically grounded ambiguity of mobile forces in northern Sierra Leone, oscillating between renovating and destroying the community and its rules, I suggest, is embodied and carried forward by young men riding motorbikes. *Okadamen* are the most visible manifestation of today's increased mobility. They also dominate Makeni's roads and crossroads where mobility is checked and balanced. *Okadamen* are thus associated with ambivalent shape-shifting forces. Therefore, their valuation was inherently indeterminate and contingent on conjunctures.

---

12 *Sathka* derives from the Arabic *sadaqa* and combines Muslim ideas of charity with non-Muslim sacrifice for the ancestors and powerful spirits. It aims to instal a social relationship with and to appease the potentially evil forces to enable oneself to govern one's own destiny (see Shaw 2002: 79–80; Jackson 1977: 32).

It is 'The Ex-Combatant' Who Rides a Bike

The history of *okada* in Sierra Leone is linked to the country's period of conflict. Nonetheless, the roots and developments of bike riding are blurry, as are the stories and histories of the war, of alliances and identities of fighters, civilians, rebels, soldiers and *Civil Defence Forces* (CDF) (see Bürge 2009). The RUF began its endeavour to liberate the population from a corrupt state, yet soon diverted from these goals and harassed civilians. In response, the CDF, originating from 'traditional' hunter societies, emerged across the country in order to defend their communities against various marauding factions. They also retaliated against civilians. The neologism 'sobel' grasps this confusion in which soldiers looted and rebels dressed in army uniforms.[13]

It was during this time of confusion that *Okadamen* emerged. The stories of their activities changed over time and according to the storytellers' position and sociocultural condition. *Okadamen* thus shared their 'destiny' with the spirits and powers of darkness (*aŋ-sum*) in Temneland who transformed from being perceived as good to bad. People accounted variably for *okada*'s roots and selectively referred to *Okadamen*'s pasts as fighters. In narratives on mobility, accidents and development, the multifaceted identities *of Okadamen* were often reduced to one determining aspect: 'They are still rebels. They have their inner habits, they don't respect others' rights and lives', a journalist explained to me in Makeni in 2007 with regards to *Okadamen's* involvement in accidents and violent demonstrations.

It is true that *okada* boomed during the conflict. In the late 1990s, a considerable number of *Okadamen* were close to or even had been part of one of the fighting factions (see Bürge 2009, 2011). Having been part of a faction, though, barely accounts for the extent and reasons for individual *Okadamen*'s involvement in the war. Furthermore, it does not explain their current visions and ideas. *Okada* had in fact perhaps been the most successful yet the least acknowledged factor in the 'reconciliation' programme. It emerged in an organic manner, without any official programmes put forward by the 'reconciliation industry'. Fighters and other young men otherwise prone to violence worked more or less peacefully together, blurring the allegedly exclusive boundaries between fighters and civilians, as they had already done during the war (Bürge and Peters 2010).

13 See Ferme and Hoffman (2004) on shifting identities of the factions and on the CDF in particular. See Peters (2006, 2011a) on the RUF, and Richards (1996) and Gberie (2005) for an overview of the conflict in Sierra Leone. Bolten (2008, 2012b) and Bürge (2009) offer accounts of Makeni's history during the conflict in which they also take issue with mainstream periodization that differed from locals' perception.

Everybody in Makeni knew at least one young man who was earning his money with bike riding and who had not been an ex-combatant. People nonetheless claimed that there was continuity between former fighters and current riders, which was noticeable in *Okadamen*'s 'rebel habit'. I argue that one reason for this was the more general connotations that 'rebel' carries. People likened them to troublesome children, which they also called 'rebels' as they lived a 'bush life' and were not (yet) fully domesticated (see Ferme 2001; Hoffman 2003; Shepler 2014). More important were the general qualities that *Okadamen* appeared to share with former fighters and the warriors who were highly ambiguous figures themselves, open to diverging selective and discriminating interpretations. As Bastian (1998: 128) has argued for the market town of Onitsha in Nigeria, 'moral commentaries in such a climate must necessarily be oblique, metaphoric, pointing to the general ambiance rather than naming specific names'. *Okadamen* embodied the general traits of possibly threatening people who could be blamed for social malaise. Therefore in people's minds and narratives 'it is "the ex-combatant" who rides a bike' (Bolten 2012a: 505).

Fixing the One with Many Manifestations

Due to the above-mentioned complexities and confusions during the war, 'combatants' bore an ambiguous reputation. Until 2013, people in Makeni continued to discuss whether members of the RUF, the *Kamajor* (CDF), the Sierra Leone Army soldiers or the ECOMOG troops had been more wicked, and which faction had been beneficial. Catherine Bolten (2012a: 500) suggests that rebels were valued in varying ways depending on their motivations for joining fighting factions. My interlocutors also distinguished between different rebels and between their origins and motivations whilst even disagreeing about these valuations among each other. Thus the category of the 'rebel' was essentially ambiguous, as were the distinctions between 'rebel', 'soldier' and 'militia' and, ultimately, between 'good' and 'bad'.

Until 2013, this confusion between the categories determined the evaluation of their alleged 'successors', the *Okadamen*. *Okadamen* have easily been identified as direct descendants of the fighting forces. Similar to combatants in the past, *Okadamen* moved quickly on the roads, appearing and disappearing as they pleased. They controlled strategic points and erected roadblocks, thus interrupting movement and circulation. They worked during the night and outside the safe realm of the town; and they made money by doing physical and often violent work. They harmed others, yet they also worked for the benefit of others. Using motorbikes, *Okadamen* not only followed demarcated and cleared roads which eased motion, but also explored new paths. Riding on bush paths and hidden tracks, *Okadamen* ventured new ways of securing social reproduction on the road. In so doing, they transformed local spaces and cosmology, and challenged the *status quo. Okadamen* adapted the protective and empowering practices of those who aided the community in the past to current exigencies. They saw themselves in continuity with these beneficial and socially productive actors. Their critics evoked another analogy, claiming that they altered social relations of control and guidance, opening the community to malevolent forces and thus endangering social reproduction. Hence, instead of protecting and helping the community, it was claimed that these young men needed to be disciplined and contained by the community.

The anxieties and aims that people verbalised were not uniform, but differed gradually. Some experienced the community and its values as truly endangered in these times of social change. Others who feared and criticised *Okadamen*'s disintegrating potential explicitly stated that it was the country's elite and its critical exposure to the world economy that had caused their 'everyday suffering' ('*ɔl de wi de sɔfa*'). Nevertheless, out of a deeply existential fear of disorder and continuous suffering, people attacked *Okadamen*. The latter were seen to be the most tangible culprits of the negative impacts of today's global economic forces, as they were the most apparent markers of local socio-economic hardship. They embodied and evoked the non-discursive memories of past confusions caused by the country's progressing regional and global integration, including the slave trade, war and the exploitation of minerals and other resources. People whose (social) mobility was rather limited and who feared being left behind tried to tie down those who trespassed thresholds, exposing and endangering themselves and ordinary people's life-worlds. These people often deeply believed that riders had to be guided and brought back onto the right path for the well-being of the community and, not least, of the riders themselves.

Others, particularly local elites (such as landowners, bike owners, NGO workers, politicians), feared a more concrete personal loss of control, as Krijn Peters (2006) has argued for immediate postwar Sierra Leone. They had an intimate interest in controlling individual *Okadamen*. For them, young men were more lucrative as dependent miners and thugs instead of working as independent self-employed and empowered riders. By evoking a genealogy of harmful 'mobile people' and by appealing to anxieties, these local elites denigrated *Okadamen*'s moral reputation. Ultimately, this led to *Okadamen*'s social marginalisation and limited their economic success.

Referring negatively to crossroads enabled people to grasp the country's problems in familiar idioms and cosmologies and to locate the alleged causes within a restricted moral landscape (see also Bastian 1998: 127). Posited as an 'immediate enemy' (Foucault 1983: 211), *Okadamen* could be made scapegoats, and punished. Problems persisted and were acerbated as only symptoms were targeted, not the causes.

Revitalising or Destroying the Community?

Okadamen tried to defy attempts of seizure, or of being 'pulled down'. From its emergence, *okada* was envisioned as a way to escape patrimonial capture and youth unemployment (Peters 2007; Bürge 2009). After the war, *Disarmament, Demobilisation and Reintegration* programs had projected ex-combatants to demobilise and settle down. *Okadamen,* combatants or not, (re)mobilised and tried uncommon and presumably uncontrollable ways to earn a living. They did not wait for their turn to become adults and to be fully integrated into society under the elders' guidance, while sacrificing their (labour) force, as coming-of-age is 'traditionally' envisioned (Shepler 2014). They attempted to realise their adulthood differently. Older people whose social mobility had been arrested due to socio-economic hardship were concerned about these successful young men's social mobility. They feared neglect and exclusion from the *Okadamen*'s networks of reciprocity. The *Okadamen* had no obligation to pay them back for care and guidance when growing up. And they feared that they would become dependants of the *Okadamen* who, in their own eyes, were in need of moral and developmental guidance. The inter-generational contract was turned upside-down. For the people who had only advanced in terms of age but had never become socially elders, *okada* was an activity of mobility that exceeded their productive control and was thus immoral.

In Sierra Leone, mobility and leaving the community had not always been seen in a negative light. In fact, people in the region had always had to move and to be active, '[f]or well-being has always been [...] a matter of bringing the vital forces of the wild into the precinct of the village – which means transgressing the symbolic boundaries that separate the secure space of home from the unknown and the beyond' (Jackson 2011: 44). Through their journeys into the wilderness, people were able to acquire new and extra-ordinary powers from outside the village or town, although this bore the risk of endangering oneself and the community. Challenging the outside was mainly the task of hunters, warriors and diviners. They opened up the community to the renewing powers of the forest, and they were able to provide protection from its dangers (see Jackson 2011: 44; Ferme 2001: 26–30). Later the concept of going out to acquire powers informed the ideal of emigrating for the purpose of furthering one's education with which one would then come back to serve the community, or for supporting those who remained back home. Yet for most Sierra Leoneans this way of relating oneself and the community to the outside powers was not often viable. Education did not necessarily lead to a well-paid job and those going to Freetown or abroad were not always able to satisfy people back home who consequently felt betrayed (Jackson 2004: 146–149).

Okadamen showed their ability to open up new ways for capturing powers and overcoming stagnation. Young men like Patrick, a shop owner at *Opin Ay* and former *Okadaman*, regularly disappeared from Makeni. Patrick went to Freetown and Conakry, sold produce from Makeni and imported items from the 'better world' outside. He did business in the bush and elsewhere, bought diamonds, gold, timber, palmoil and cannabis. While working for an international construction company, he often got the opportunity to make a 'secondary' business in the hidden. Patrick brought fancy goods to Makeni and shared his money with friends. In 2007 he had stories to tell about 'enlightened' Conakry (where electricity was more stable than in any place in Sierra Leone) and about the 'European life' he had lived there, including drinking cold drinks and going to clubs. For many of his listeners, Patrick's stories and imported goods served as kinds of revitalising forces. His economic success was tangible proof of a better life that was possible and attainable, but which usually only existed in images, on television and in people's dreams. The point was that one had to leave Makeni in order to capture these 'forces'.

Entering the outside at the crossroads where the worlds meet requires securing this threshold and repulsing or controlling the powers that spill over. Many people in Makeni longed for imported goods that made life easier, whereas others rejected them for their unknown origins and powers. These powers potentially threatened the community, as was the case with roads. Nobody had ever seen how and out of what most of the roads were exactly made. 'White men' and 'Chinese people' produced them in faraway factories. Black people could work in these factories, in the USA for example. Yet allegedly no 'black man' had access to the central production unit where unknown powerful actors made inventions. The terms of production and acquisition of foreign goods were not clear, and neither were their long-term impacts. Improved roads and the vehicles on them accelerated transport, mobile phones eased up communication and financial transactions, electricity made life more comfortable – yet only for those able to control these technological 'improvements'.

When Patrick disappeared into the bush during the night, it remained unclear where the powerful, expensive and yet unknown items he brought back came from. Assumedly they stemmed from nebulous business with unknown forces that he encountered on his journeys into other worlds beyond the visible one (*nɔ-ru*) and beyond the safety of the community. In the 'underworld', as people called the worlds beyond the visible, human beings were sacrificed for gaining power. It was thus assumed that Patrick made business in anti-social ways, endangering instead of protecting the community. Bluffing with his wealth in front of his friends made him into a cunning businessman. In other instances, he changed his appearance and pretended to '*drɛg*' (survive) as a poor *Okadaman*. His periods of absence became increasingly longer and a great part of his acquired wealth withered away just as it had come. This confirmed people's view of him being someone who had little in common with protective hunters and warriors who opened up the community for renewing forces. Rather, Patrick and other *Okadamen* resembled the 'witches and other transmogrifying creatures' Ferme (2001: 26) refers to. These forces populate the realms of darkness (*aŋ-sum*), the bush and the night; they shift appearance as they move by night between different places and endanger the community. By using and controlling roads and rivers for their purposes, *Okadamen*'s activities conjured up the harmful powers that had exploited the country for centuries.

Young men tried to navigate difficult conditions through the possibilities that bike riding offered. Yet beyond mere survival, they also wanted to enjoy life. The younger ones among them especially wanted to go out, to drink and to impress peers (*blɔf*) and pick up (*cajole*) girls. They enjoyed the speed as well as the freedom from school or parental control. More experienced riders tried to send their children to school and saved money in order to settle around the junction and open a business there, particularly as they felt that riding exhausted their bodies. Offering valuable services to the community, they expected gratitude and possibilities for upward social mobility. Around the junction *Okadamen* occupied various niches in order to remain active, not waiting for but enforcing what Charles Piot (2010: 66) calls an 'intercession that might be life-transforming'. Flexibility, diversified loyalties and the mediation of different realms distinguished *Okadamen* – qualities which had also been appreciated in Makeni during other eras. Excessive social criticism, however, urged many riders to limit their engagements at the crossroads. They started forming networks in other places, applying tactics of hiding and dissimulation. Ultimately, it was not motorbike riding that made *Okadamen* neglect their peers and that made their economic endeavours 'unsustainable', but people's mistrust and envy (*bad at*).

Conclusion

Okadamen have often been made scapegoats for Makeni's problems and pushed to the margins of society. As I have argued in this chapter, these ascriptions of qualities to people were selective, but not merely arbitrary 'fixations' of their ambiguous valence. They were re-enactments of memories of ambivalent forces inscribed in crossroads and embodied by the young men who gather at them. I have shown that both, crossroads and *Okadamen,* could evoke or be linked to divergent memories. A road always leads into two directions, literally and metaphorically. Roads can be stairways to prosperity or to ruin. They are the channels on which empowering and rejuvenating goods and people are brought to a place or on which the powers and wealth of the place are carried away. Roads connect to the wider world and to the dreams linked to those far-off places.

While people inform the identity and meaning that are ascribed to roads, roads inform the identity of the people who use them. In Makeni, motorbike riders were the most visible actors entangled with roads. Due to their intimate relation with the likewise ambiguous roads, *Okadamen* were considered to be rife with ambivalent forces themselves. *Okadamen* were ambivalent because they resembled past actors with similar positions on the road and at the crossroads. They were mobile, as were the fighting forces in the past. They controlled crossroads and other check-points as rebels had done. And like hunters, they dared the darkness and left the trodden paths. The qualification of *Okadamen*'s power was thus indeterminate and open to people's selective assessment. The crossroads offered a stage for public discussion, accusation and meditation of people's qualities. At the same time, crossroads co-determined these assessments due to the interrelated production of the identity of related actors. In other words, the ambivalent ways in which people perceived *Okadamen* were intimately wedded to the ways in which people perceived the ambivalent nature of crossroads.

The mutually constituting – human and non-human – powers of crossroads provided what can be seen as 'moral frameworks for meditating on the relation between individuals and society, the origins of value, morality and the production of wealth' (Sanders 2008: 122). At crossroads people tried to make sense of the world that they inhabited. Their diagnosis of Makeni's contemporary condition was shaped by the legacies of ambiguous experiences of the past with controversial actors and activities, present exigencies and longings for the future, all of which was condensed at the crossroads. Young men riding motorbikes could be discerned as 'makers or breakers of society' (Honwana and De Boeck 2005); their practices could lead to a better future, but it could also threaten social reproduction.

When commercial motorbike riding emerged, the general public appreciated the flexibility and mobility of *Okadamen* not only for their transport services. *Okadas* allowed 'civilians' and 'combatants' to mingle in-between supposedly clear-cut 'identities', contributing to what is often called *re-integration* even before the end of the war was officially declared. This mingling of people standing in-between identities had cohesive effects on life in Makeni in the late 1990s and early 2000s. Bolten (2008) has shown how this blurring of clear distinctions was later used politically by the post-war SLPP government. The RUF 'stronghold' Makeni was marginalised within Sierra Leone and cut off from social and infrastructural re-construction. Ironically, the stigmatisation of those who were thought to be congruent with rebels was reproduced in a different manner during my research. People reduced *Okadamen*'s flexible qualities and fragmented histories to one negative identity that equated them with ex-combatants and evil spirits. *Okadamen* were fixed as *the* evil, as was Esu Elegbara: the Yoruba trickster-god of the crossroads known as 'the one with many manifestations […] who straddles all realms and acts as an essential factor in any attempt to resolve the conflicts between contrasting but coterminous forces in the world' (Aiyejina 2009: 3–4).

The roles of the various powers for the community's well-being have been ambivalent throughout time. They depended on conjunctures, on the particular ways in which powers come together and affect the positions of the individuals at a certain point and time. People's opinions about the impacts of these powers have differed. Conjunctures will continue to change, and so will the configuration of the powers at the crossroads.

References

Achebe, Chinua. 1975. *Morning Yet on Creation Day*. Garden City: Anchor Press.

Aiyejina, Funso. 2009. *Esu Elegbara: A Source of an Alter/Native Theory of African Literature and Criticism*. Distinguished Lecture Series, Centre for Black and African Arts and Civilisation. Lagos: Obafemi Awolowo University.

Bakare-Yusuf, Bibi and Weate Jeremy. 2005. 'Ojuelegba: The Sacred Profanities of a West African Crossroad.' In *Urbanization and African cultures*, edited by Falola Toyin and Salm Steven J., 323–341. Durham: Carolina Academic Press.

Bakhtin, Mikhail. 1981. *The Dialogic Imagination*. Austin: University of Texas Press.

Bastian, Misty. 1998. 'Fires, Tricksters and Poisoned Medicines: Popular Cultures of Rumor in Onitsha, Nigeria and Its Markets.' *Etnofoor* 11 (2): 111–132.

Bolten, Catherine E. 2008. *'The Place is so Backward': Durable Morality and Creative Development in Northern Sierra Leone*. PhD thesis, University of Michigan.

Bolten, Catherine E. 2012a. '"We Have Been Sensitized": Ex-Combatants, Marginalization, and Youth in Postwar Sierra Leone.' *American Anthropologist* 114 (3): 496–508.

Bolten, Catherine E. 2012b. *I Did it to Save My Life: Love and Survival in Sierra Leone*. Berkeley: University of California Press.

Bürge, Michael. 2009. *'Mal geht es bergauf, mal geht es bergab!' Erfahrungen von Motorrad-Taxifahrern in Sierra Leone*. MA thesis, University of Zurich.

Bürge, Michael. 2011. 'Riding the Narrow Tracks of Moral Life: Commercial Motorbike Riders in Makeni, Sierra Leone.' *Africa Today* 58 (2): 59–95.

Bürge, Michael and Peters Krijn. 2010. 'Soziales Kapital als Potenzial kollektiver Demobilisierung. Das Beispiel der Motorrad-Taxifahrer in Sierra Leone.' In *Jugendliche in gewaltsamen Lebenswelten: Wege aus den Kreisläufen der Gewalt*, edited by Kurtenbach Sabine, Blumör Rüdiger and Huhn Sebastian, 163–176. Baden-Baden: Nomos.

Christensen, Maya M. and Utas Mats 2008. 'Mercenaries of Democracy: The 'Politricks' of Remobilised Combatants in the 2007 Presidential Elections, Sierra Leone.' *African Affairs* 107 (429): 515–539.

Dalakoglou, Dimitris and Harvey Penny. 2012. 'Roads and Anthropology: Ethnographic Perspectives on Space, Time and (Im)Mobility.' *Mobilities* 7 (4): 459–465.

Euba, Femi. 1996. *Camwood at Crossroads*. N.p.: Xlibris.

Fanthorpe Richard and Roy Maconachie. 2010. 'Beyond the 'Crisis of Youth'? Mining, Farming, and Civil Society in Post-war Sierra Leone.' *African Affairs* 109 (435): 251–272.

Ferme, Mariane. 2001. *The Underneath of Things: Violence, History, and the Everyday in Sierra Leone*. Berkeley: University of California Press.

Ferme, Mariane and Hoffman Daniel. 2004. 'Hunter Militias and the International Human Rights Discourse in Sierra Leone and Beyond.' *Africa Today* 50 (4): 73–95.

Foucault, Michel. 1983. 'The Subject and Power'. In *Michel Foucault: Beyond Structuralism and Hermeneutics*, edited by Hubert L. Dreyfus and Paul Rabinow, 208–226. Chicago: University of Chicago Press.

Gberie, Lansana. 2005. *A Dirty War in West Africa: The RUF and the Destruction of Sierra Leone*. Bloomington: Indiana University Press.

Green-Simms, Lindsey. 2010. '"No Danger No Delay": Wole Soyinka and the Perils of the Road.' *Journal of Postcolonial Writing* 46 (1): 53–64.

Hansen, Karen T. 2005. 'Getting Stuck in the Compound: Some Odds Against Social Adulthood in Lusaka, Zambia.' *Africa Today* 51 (4): 3–16.

Hoffman, Daniel. 2003. 'Like Beasts in the Bush: Synonyms of Childhood and Youth in Sierra Leone.' *Postcolonial Studies* 6 (3): 295–308.

Hoffman, Daniel. 2011a. *The War Machines: Young Men and Violence in Sierra Leone and Liberia*. Durham: Duke University Press.

Hoffman, Daniel. 2011b. 'Violence Just in Time: War and Work in Contemporary West Africa.' *Cultural Anthropology* 26 (1): 34–57.

Honwana, Alcinda M. 2012. *The Time of Youth: Work, Social Change, and Politics in Africa*. West Hartford: Kumarian Press.

Honwana, Alcinda M. and de Boeck Filip (eds). 2005. *Makers & Breakers: Children & Youth in Postcolonial Africa*. Oxford: James Currey.

Ingold, Tim. 2011. *Being Alive: Essays on Movement, Knowledge and Description*. London: Routledge.

Jackson, Michael. 1977. *The Kuranko: Dimensions of Social Reality in a West African Society*. London: C. Hurst.

Jackson, Michael. 1990. 'The Man Who Could Turn into an Elephant: Shape-Shifting among the Kuranko of Sierra Leone.' In *Personhood and Agency: The Experience of Self and Other in African Cultures*, edited by Jackson Michael and Karp Ivan, 59–78. Uppsala: Uppsala Studies in Cultural Anthropology.

Jackson, Michael. 2004. *In Sierra Leone*. Durham and London: Duke University Press.

Jackson, Michael. 2011. *Life Within Limits: Well-Being in a World of Want*. Durham and London: Duke University Press.

Kanu, Alhassan Fouard. 2012. 'An 'Okadised' Society: Implications on Teenage Pregnancy in Sierra Leone.' *Concord Times* (February 1). Accessed July 18, 2016. http://allafrica.com/stories/201202010784.html.

Klaeger, Gabriel. 2012. 'Rush and Relax: The Rhythms and Speeds of Touting Perishable Products on a Ghanaian Roadside.' *Mobilities* 7 (4): 537–554.

Klaeger, Gabriel. 2013. 'Introduction: The Perils and Possibilities of African Roads.' *Africa* 83 (3): 359–366.

Lamont, Mark. 2013. 'Speed Governors: Road Safety and Infrastructural Overload in Post-Colonial Kenya, c. 1963–2013.' *Africa* 83 (3): 367–384.

Littlejohn, James. 1963. 'Temne Space.' *Anthropological Quarterly* 36 (1): 1–17.

Masquelier, Adeline. 2002. 'Road Mythographies: Space, Mobility, and the Historical Imagination in Postcolonial Niger.' *American Ethnologist* 29 (4): 829–856.

Masquelier, Adeline. 2013. 'Teatime: Boredom and the Temporalities of Young Men in Niger.' *Africa Today* 83 (3): 470–491.

Menzel, Anne. 2011. 'Between Ex-Combatization and Opportunities for Peace: The Double-Edged Qualities of Motorcycle-Taxi Driving in Urban Postwar Sierra Leone.' *Africa Today* 58 (2): 97–127.

Peters, Krijn. 2006. *Footpaths to Reintegration: Armed Conflict, Youth and the Rural Crisis in Sierra Leone*. PhD thesis, Wageningen University.

Peters, Krijn. 2007. 'From Weapons to Wheels: Young Sierra Leonean Ex-Combatants Become Motorbike Taxi-Riders.' *Journal of Peace, Conflict and Development* 10: 1–23.

Peters, Krijn. 2011a. *War and the Crisis of Youth in Sierra Leone*. Cambridge: Cambridge University Press.

Peters, Krijn. 2011b. 'The Crisis of Youth in Post-War Sierra Leone: Problem Solved?' *Africa Today* 58 (2): 128–153.

Peters, Krijn and Richards Paul. 1998. 'Why we Fight: Voices of Under-Age Youth Combatants in Sierra Leone.' *Africa* 68 (2): 183–210.

Piot, Charles. 2010. *Nostalgia for the Future: West Africa after the Cold War*. Chicago: University of Chicago Press.

Richards, Paul. 1996. *Fighting for the Rain Forest: War, Youth and Resources in Sierra Leone*. Oxford: James Currey.

Sanders, Todd. 2008. 'Buses in Bongoland: Seductive Analytics and the Occult.' *Anthropological Theory* 8 (2): 107–132.

Shaw, Rosalind. 2002. *Memories of the Slave Trade: Ritual and the Historical Imagination in Sierra Leone*. Chicago: University of Chicago Press.

Shepler, Susan. 2014. *Childhood Deployed: Remaking Child Soldiers in Sierra Leone*. New York: New York University Press.

Simone, AbdouMaliq. 2005. 'Urban Circulation and the Everyday Politics of African Urban Youth: The Case of Douala, Cameroon.' *International Journal of Urban and Regional Research* 29 (3): 516–532.

Stoller, Paul. 2002. 'Crossroads: Tracing African Paths on New York City Streets.' *Ethnography* 3 (1): 35–62.

Wylie, Kenneth C. 1973. 'The Slave Trade in Nineteenth Century Temneland and the British Sphere of Influence.' *Africa Studies Review* 16 (2): 203–217.

CHAPTER 7

Ruin, or Repair?

Infrastructural Sociality and an Economy of Disappearances along a Rural Road in Kenya

Mark Lamont

Max Gluckman's (1940) 'Analysis of a Social Situation in Modern Zululand' is arguably one of anthropology's most celebrated essays on the social and historical conveyances of power in colonial Africa. Through a deftly wrought 'apt description' of the hustle-and-bustle attending to a secular ritual of a bridge opening ceremony in the Zulu hinterland, Gluckman's portrayal of pre-apartheid rural South Africa made a lasting analysis of the social complexities of colonial relationships. Although this essay's achievement was quite materially built up around works of engineering, both political and infrastructural, it remained curiously silent on the situated sociality of infrastructure in the ritual's enactments as state effects. Roads, bridges and automobiles were seemingly passive props in this localised dramaturgy, while the relationships framed and rendered explicit, and thus changed, were mediated by exactly these vital integrations of infrastructure and technology. While we cannot with any satisfaction accuse Gluckman of having 'missed something' in his preliminary analysis, we can ask: Whatever happened to the structure built across the Malungwana Drift in 1938? Where is this bridge now?

No irony is lost on the probability that the literary status of 'the Bridge' essay far out-lives the material, historical bridge itself. If this bridge formed part of what Brian Larkin (2008) called the 'colonial sublime', that sense of enchantment that ties technology to alluring promise, how can we theoretically account for how this grasp of people's dreams fades away in time? How can anthropological interests in the seemingly 'obdurate' qualities of infrastructure think about the inevitability of its fatigue and decay? How can we develop methods to follow the shine and polish of the new to the patina and rust of the old?

These questions echo a collection of recent writings addressing the temporality of infrastructure and the limited life-spans of technologies, particularly with respect to the global expansion of road building under neo-liberal conditions (Harvey and Knox 2012; Hetherington 2014). Grounded in ethnographies from the depths of Amazonia to inner Mongolia, these writings foreground how roads, in particular, do much more than convey people and goods across

 | DOI 10.1163/9789004339040_008

a political and economic landscape, but that infrastructure – in its efficacy – enables and enacts myriad kinds of 'enchantment' and 'affective engagements' with respect to the promises of speed, political integration and economic connectivity (Harvey and Knox 2012: 534). Seen in light of these ethnographic writings on roads and associated infrastructure, the promissory nature of state projects is revealed, alongside the temporal fragility of infrastructure's materiality and its political promises. Such authors highlight how the material processes of road building 'calls forth competing, unauthorised and openly unstable dimensions of being' (Harvey and Knox 2012: 521), such as the shifting seasonal qualities of road surfaces or political exigencies that confront and interrogate these promises. Focusing upon 'enchantment' as a way to think of infrastructure's sublime and figural presence within everyday life, several ethnographers have brought to our attention the underlying morality of road making in being far from equivocal as a 'public good', but also facilitating the establishment of a variety of moods or affects.

This focus on the 'enchantment of infrastructure' sets the tone of debate for this chapter, but the road described here parts company with figural visions of freshly carpeted, sleek and paved highways. Instead, enchantment operates in a reverse logic: as when the ethnographer finds in such infrastructural decay a kind of forensic wonder, asking how such road surfaces came to be so rotten.

The bridge at the Malungwana Drift has likely gone the way of the silt of the Black Umfolozi river, being slowly washed away over the decades. We know that the shifting sands of the watercourse, as well as the porosity of aged and decaying concrete, are material realities, but we are less prepared to see in these groaning conditions certain kinds of state effects themselves. Indeed, through reflecting on recent anthropological writings on the materialisation of state power through infrastructure, we might forget that part of the bedazzle of this particular show of state power is conveyed in the 'delivery' of development of these kinds – roads, bridges, reservoirs, telecommunications masts – captured in the banal work of ribbon cutting rituals. What if we were to build the passage (and ravages) of time into these considerations of infrastructural power, such that the erosions and erasures that fixed infrastructures eventually endure come to be seen as state effects? What if we were to consider state power in the provision and regulation of routine maintenance, if not in the laissez-faire neglect of projects once 'delivered' on short-term political promises?

Such questions frame the theoretical argument being pursued in this contribution on roads and associated infrastructure in rural Kenya. I am chiefly interested in examining the ways in which people must relate to each other when infrastructure decays and how this reflects both state effects and various forms of state and non-state power. The ethnographic framework is not about the building of a new road, about 'makings' in this sense, but rather stems from an analysis that is forensic in more than one way.

Firstly, the ethnography goes back to a time when the very idea of a new road generated local frictions and tensions in an 'economy of disappearances', to subject Anna Tsing's useful phrase, economy of appearances, to a context specific *détournement* (Tsing 2005). This is, in short, the ruined agrarian economy based on the small-holder production of coffee and tea and its co-operative marketing regimes, crashed-and-burned by the neo-capitalist reforms of the late 1990s. In this, I aim to look at how the decay and destruction of infrastructure in Kenya can also enchant and, thereby, provoke affective engagements with the very materialities of this ruination itself. I examine the usages and repair of roads, bridges and culverts in the absence of any clear state provision of maintenance in order to convey the politics set in motion (and halted to a standstill) by the economy of disappearances.

Secondly, I show that infrastructure can also 'disenchant' by delving into the affective dimensions of disappointment, exasperation and loss that attends to failures to deliver on political promises (see Dalakoglou and Harvey 2012). I reflect on this as the 'forensic wonder' evoked by engagements with ruins that trace out of the presence of the past in the current landscape (see also Gordillo 2014 on rubble). This questioning by those confronted by the ruined condition of their roads and associated infrastructure can add further conversation to anthropologists writing about the embodiment of 'affect' in human relations to materiality, particular those of nostalgia and melancholia (Navaro-Yashin 2009), but also those touching upon landscapes experienced through aspiration and hope (Miyazaki 2004; Fontein 2011). With respect to ethnographies of the global post-colony, attending to questions of materiality and 'affective infrastructure' (see Street 2012) revisits the 'ethnography of decline' envisioned by James Ferguson (1999), bringing back into focus his insightful argument that modernity is an incomplete frame, as when 'linear teleologies of emergence and development remains an unfinished task' (Ferguson 1999: 17). For my more parochial concerns along the line of road in rural Kenya, this return to a historical perspective also bring home the salience of Sara Berry's (1993) aphorism that 'no condition is permanent' (Berry 1993), seen in the cadastral ghosts of not-so-old survey maps used as toilet paper, and lumps of tarmac slag, seeded with vegetation at the road side.

More recently, an ethnography of decline is emergent (see Tsing 2015; Gordillo 2014), looking at all forms of ruin and rubble, one outcome being theoretical insights into how new forms of sociality spring forth from the debris left behind by cycles of destruction and violence. Ruined infrastructure commands our attention, not only because ordinary social life is disrupted, but also because people find creative ways of mending, resolving or redirecting such disruptions. It is this creative interplay between everyday life being halted by ruined infrastructure and making something new out of its debris that I identify as infrastructural sociality. By infrastructural sociality, I mean those instances where people form relations of necessity in resolving problems associated with infrastructure. The idea of necessity is basic to the *modus vivendi* that people must form with one another, through the agency of things, when the circulation of the most essential materials for everyday life are interrupted altogether. Breaches in the working order of most infrastructure brings into sharp focus the function and form of our social inter-dependency, but we examine less frequently how disruption creates new kinds of social relations and values.

Infrastructural sociality is implied in the most banal and essential everyday tasks, but here, I have in mind the ways in which the accidental occurrence of infrastructural decay or breakdown leads people to explore and expand previously unknown relational possibilities. Everyday examples include the kinds of rapport that passengers strike up with one another while travelling in public transport vehicles, such as either acquiescing to hiked fares or protesting against them; or otherwise, dismounting from a vehicle while passing a particularly muddy stretch of road, even pulling together to free buses from ditches or muddy ruts, their exertions inseparable from those of the vehicle's crew. Alternatively, it also has the sense of people maintaining infrastructure, such as bridges, on their own accord through the failure of the state to provide the labour, resources and equipment to bring about repairs or maintenance. It could also be extended further to all manner of everyday repairs made to plastic basins, radios and water pipes, among other everyday material things.

Picking up the story just before massive investments in road infrastructure transformed (auto)mobility throughout Kenya, I delve into reflexive questions asked by locals along a particular rural road in Kenya to elucidate the socially generative qualities of infrastructural ruin and repair. In short, I have sought to discuss these qualities by focusing on the subtle reach of infrastructural power through a Hamlet-like musing: 'to ruin or not to repair, that is the question'. In bringing these two possibilities of power to light – in the dialectics of destruction and construction – I thus address a far more subtle, even ignored aspect of everyday life that turns on the question of how certain kinds of infrastructure, roads and bridges in particular, come to shape the very sociality of their usage. I offer in this chapter an examination of the makings of new kinds of social relations through differing kinds of engagements with infrastructural maintenance or neglect along Kenya's rural road network.

Return to the Economy of Disappearances

In August 2008, I was travelling back to one of my former research sites in a taxi through a maze of temporary access roads, built as detours whilst extensive road engineering works were in progress. Besides the apprehensions that any anthropologist might face on returns to the field, I was particularly excited by the prospect of seeing the Meru-Mikinduri-Maua Road finally being ripped up, realigned and rerouted by a fleet of heavy machines in preparation of being paved, an act of government long anticipated in this rural part of Kenya. It was also a moment of renewed research interests, as I was then beginning to follow what I have called elsewhere Kenya's new infrastructural dispensation and the lived state effects of President Kibaki's development plan: Vision 2030. While gingerly navigating a deep rut worn into the soil by heavy traffic, we caught a glimpse of bright yellow excavators in the far distance, antlike machines crawling through a haze kicked up by clouds of red dust, equipment slowly but forcibly reshaping the landscape through which the road cut. It was a scene being replicated in many parts of Kenya from about 2007 up to the present.

As an anthropologist, I had experienced this road as an integral part of my fieldwork five years previously. Indeed, the arduous and demanding drive along its constantly changing corrugations and washboards, its rocks and runnels, provided a visceral and mechanical grounding of 'life as lived' in this region that mirrored the 'real' work of data collection throughout the course of my research. Many memorable encounters occurred when I was behind the wheel of a decrepit old Land Rover, lovingly nicknamed *The Queen Mother*. I had no way of knowing that journeying along this road would, in time, become an object of my subsequent ethnographic writings, nor could I anticipate, in that moment, the role that infrastructure and motor vehicles would play in my theoretical understandings of everyday life in Kenya.

Although most of this road was recently paved with high-density tarmac in 2009, my story begins ethnographically in 2002 when I moved to Mikinduri from another field-site, the larger urban centre of Meru Town (Mûtindwa). With all my worldly possessions, I spent four hours travelling the short distance between these centres in a Datsun pickup truck converted into a PSV (Public Service Vehicle), better known as *matatu*, when this road was largely an impassable quagmire and in a condition that could be accurately described as being in a state of ruination. Having become stuck in the mud no less than five times, passengers and crew worked together to pull and push the vehicle free, soiling our clothes with mud and sharing the irony of the *matatu*'s logo, *Amua Twende!*, 'Agree, Let's Go!' each and every time the burdened vehicle lurched forward or slid sideways. Because of the poor transportation links between these two field-sites, the twenty kilometres that separates them could have been two hundred kilometres, particularly when comparing the unequal level of infrastructural development between Meru Town and Mikinduri. I was shocked by the transition between these two towns on my first visit to Mikinduri during the rainy season, finding its central thoroughfare a gummy tract of sticky mud, while people walked above murky pools of sitting water along make-shift duckboards of odd bricks and rotting planks. The gritty materiality of Mikinduri's line of shops with their hand-painted signs, the roughly constructed buildings and warren of kiosks and the old-fashioned government telephone exchange contrasted with Meru Town's expansive bus and taxi stages, its ATMs, tourist-class hotels, internet cafés and massive retail markets, despite the colonial patina of some parts of the town and the slums on its fringes.

This chapter places this road, officially known as D484, into both historical and ethnographic context, that is, before it was paved. This might seem a counter-intuitive approach, seeming to avoid an opportunity to speak of the conditions and aspirations or promises and perils that a new road may bring. I write forensically, when the old road lay in ruins, to make overarching theoretical points about how the political agencies of ruination and repair, as particular kinds of state effects, become visible markers of the temporality of infrastructure within the horizons of larger political-economic booms and busts. The primary example of such a formation, here, is the modernist vision of agrarian development, 'smallholder utopia' in another frame (Campbell 2012), that sustained Kenyan hopes for prosperity and opportunity from the first decades of political independence from Britain in the early 1960s to the politically angry 1990s, a decade during which I made several purposed visits to the country. In 1998 this region was reeling in the consequential aftermath of neoliberal market reforms and marked the beginnings of what I am calling the economy of disappearances, that is, the abrupt and visible decline of coffee and tea as viable mainstays of the agrarian economy (see Lamont 1999). This political-economic context of decline is critical to this essay, since the whimpering and pessimistic end to the agrarian dream that kept smallholders holding on to hope is linked, in myriad ways, to the infrastructural decay described as evidence of the economy of disappearances witnessed during periods of fieldwork. This decline and its affect of disappointment and loss of faith in the promises of development (if not the 'expectations of modernity') also underlay the figural and ideological qualities of political promise in the region, particularly of greater connectivity through the building of sleek surfaced and fast running paved roads.

I am largely interested in getting readers to share in some of these affects, particularly the sense of being temporarily stuck or stranded as a driver or passenger along D484 and its rhizomic offshoots, optimistically named 'access-roads' by remote bureaucrats in Nairobi office blocks. I want readers to participate imaginatively in the problem-solving and collective actions that broken infrastructure compels. By drawing attention to the ways in which infrastructural breakdown provokes specific kinds of repair or neglect, my notion of infrastructural sociality rests upon the observation that when a road's surface is so poorly maintained that it is dangerous and unpredictable, hence slowing or even stopping any motorised travel over it, it demands attention through its very haphazard materialities. In other words, as Stephen Graham and Nigel Thrift (2007: 3) argue for repair and maintenance in other contexts, the 'breakdown' of infrastructure's working order – beyond just making it 'visible' – has a secondary, yet vitally important effect of also 'abruptly' demanding 'interaction with itself' (Verbeek 2004: 79–80, quoted in Graham and Thrift 2007: 3). This is key to understanding the particular relations and sociality that took place along impassable points in the road: as when vehicles sank immobile into muddy ruts and all able-bodied passengers and onlookers pulled together to muscle it free; or to provide another example, as when people came together to lift fallen power cables out of the road's reach, engaging creatively and dangerously with such wires using a series of rope pulleys to reposition them to a suitably 'safe' height by hooking them over road-side tree branches. It is this 'demand' of common, needful materiality coupled with the forensic wonder of discussing how something came to be as disorderly and challenging to work with as it was that I am flagging as 'infrastructural' sociality.

These theoretical concerns blend with considerations of method. It was this kind of participant-observation with the unpredictable and volatile materialities of travel along D484 that put me in the driver's seat, so to speak, of many of the issues raised here. Like Jeremy Campbell's (2012) ethnography of seasonal travel along an Amazonian highway, the heterogeneity of broken and viscous road surfaces of D484 also served to provide 'sense', in so far as travelers and by-standers alike entered into engaged conversations about the connections between frustrating and dangerous materialities and the state's promissory character. Although the landscape through which this road ran was littered with ruins of past engineering and cadastral planning – the economy of disappearance being only the latest such ruination – travel through the hilly environment by four-wheel-drive or lorry was experienced as a series of protracted stops and starts (see Kernaghan 2012), depending on the surface condition of the road itself and the shifting material integrity of culverts, telephone poles and power cables and bridges.

Since driving is as much a part of fieldwork as walking (or any other activity), here I propose that this relatively short stretch of road (about 52 kilometres from Meru to Maua) can tell us a lot about the social aspirations and frustrations of those people whose everyday lives are shaped by the materiality of this particular road and the (dis)connectivity it makes possible. I do not intend to ignore the place of this road in the wider context of Kenya, especially Kenya as experienced through access to tarmac roads, but I would wish to make the argument that the ideology implicit with the design and function of infrastructure is not only hidden from view when 'things' are in working order (Graham 2010; Humphrey 2005; Simone 2004), but that its very breakdown forces people to confront the materiality in itself, as well as its sense and its figural qualities (see Campbell 2012; Kernaghan 2012). Neither can this methodological focus on motorised, but visceral movement and travel – even 'wayfaring' (Ingold 2000) – ignore the significance of the landscapes in which such failing infrastructure is found, in its layered qualities and the longer historical narratives through which people either commemorate or silence how the 'road was built' (Roseman 1996). Writing about the road *as if* it has always been paved and maintained to good working order would cover over much of what constituted the road's social and political sense, that is, to ignore its forensic dimension and its capacity to give voice to the erosions and erasures in the landscape that made its own present conditions possible. As such D484's transformation from ruination to renovation in the space of a decade can be a window onto the national politics of Kenya's recent 'boom' in infrastructural investment, particularly through road construction, but more significantly it can reflect the temporalities of such infrastructural investments and beg further questions about the future politics of maintenance.

In what follows, I discuss the recent histories of mobility and infrastructural development in Meru through examining national politics from the perspective of my hosts in Mikinduri, Tigania. While I have not been witness to the last decade in Mikinduri from an ethnographic perspective, I aim to weave together a number of sources including electronic archives, on-line newspapers, blogs, emails and phone-calls to contextualise some of the changes that a paved road has brought, particularly how these transformations in infrastructure were effected by the Government under the decade-long presidency of Mwai Kibaki. Following from this, I explore in more detail the landscape through which the Meru-Mikinduri-Maua road runs, presenting ethnography about driving through this hilly terrain as a way of expanding the notion of infrastructural sociality.

D484, or Karaurî's Enigma

I have never laid eyes on Mikinduri's new paved road. When last in Mikinduri during whirlwind visits in 2007 and 2008, the large construction firm, H-Young, was undertaking a massive landscaping project, slowly bulldozing a total of 1,594,000 cubic meters of earth to clear the way for further tarmac engineering to be completed in August, 2009.

H-Young's contract, worth $27.3 million USD, was part of Mwai Kibaki's pledge to rebuild Kenya's faltering economy through commitments to improving infrastructure, principally in the form of liberalised telecommunications, road construction, port expansions, irrigation and energy generation. Improving roads and transport was one of the pledges that brought NARC (National Rainbow Coalition) into power following the 2002 elections, emerging victorious over KANU (Kenya Africa National Union). NARC's coalition of opposition parties focused on constitutional reform, but roads and transport were a major stress throughout their campaigns in both rural and urban constituencies, indelibly linking 'improved infrastructure' to poverty eradication and economic recovery.

Kenya's road infrastructure policies have undergone four rough phases in a half-century of postcolonial independence (see Wasike 2001): one initial period of rapid growth and massive borrowing for road building projects, leading to infrastructural overload and the quick deterioration of 'new' tarmac roads (1964–1972); a concomitant period of intense decline in road maintenance precipitated, in part by the international 'oil crisis', poor institutional management and antagonism between transport industry stakeholders and varying governmental bodies (1973–1982); the infamous implementation of structural adjustment and fiscal austerity, leading into the political violence and tumult of multiparty democracy in the guise of 'good governance', the effect of which was the use of road construction and repair to curry support for KANU (1982–2002); followed by a period of extensive neo-liberal reform and constitutional review since about 2002 under the watchful audits of the international financial institutes and its models of global governance where 'infrastructure' and 'mobility' are buzz-words (see Lamont 2012).

What this meant for the residents of Mikinduri and those living along D484 was a long period of neglect and government indifference. In the view of many Tigania I knew, the road and its infrastructural environs were indices of the oppositional politics of farmers in the central highlands of Kenya since the early 1990's. The very poor condition of Kenya's motorised roads in such agriculturally dependent counties (then called districts) was a fundamental political issue and focus of debate that was particularly audible during my fieldwork.

Tigania entered coffee and tea production quite late, after 1979, and is associated with the election of Mathew Adams Karaurî, Tigania East Constituency's most long-standing KANU MP. Tigania's farmers have historically faced severe problems getting their produce to markets because of the destructive force of rain on gravel roads. In this hilly landscape, transporters risk flash floods and the heavy rainfall washes away even the hardiest 'soft' road infrastructure. These links between government inaction regarding roads, the predictable ruination of road infrastructure with the rains and the shifting political alliances of local farmers, especially as their incomes declined under the Moi presidency, made this road one of the main focuses of Karaurî's political career spanning three decades.

Walking in the company of an old friend, Kobia, on my 2008 visit, we climbed above Mikinduri to the summit of a steep hill called Kilimantuune to survey the road works then underway. From this vantage point, we could see just how much earth had been removed to make way for the new road, Kobia pointing out where farmers had lost significant portions of their land to the excavators. 'We Tiganians', the forty-year-old businessman said, 'have been waiting for a paved road all my life.' In the build-up towards the 2002 General Elections, which saw the ousting of KANU politicians from all the Tigania constituencies (East and West), Kobia was adamant that Karaurî was 'finished' by his challengers come the impending elections. 'Look', Kobia asserted, 'Tigania has been *systematically* left out of major developments occurring in other parts of Meru', stressing this as a deliberate action by certain leaders.

> We want development. Karaurî had pretty much three decades to bring development, but instead we got all these impassable roads, while he was saying all along that we should fend for ourselves: if we couldn't transport our produce to market, then we should walk it there; if the Government tractors didn't come, then we should get out our *jembes* (hoes) and fix the road ourselves. No, Karaurî is finished now. KANU is finished now.

When I arrived for fieldwork in Tigania, Karaurî was a very unpopular man, but it had not always been the case. Entering parliament in 1979, Karaurî dominated Tigania politics until 2002, only being uprooted from his seat from 1992 to 1997 by the late Benjamin Ndubai. Whenever 'on the road' with local people during my fieldwork and we encountered a rough section of driving (which was most of the time), I learned about Karaurî's alleged misdeeds while he was MP for the constituency, accusations that did not always accord with the records of his time in Parliament housed in the official government Hansard. D484 had been a priority and, indeed, roads a major preoccupation of this politician.

In particular, Karaurî was to raise the question of repairing and rebuilding the bridge over the river Thangatha, a watercourse that shrinks to a stream during the dry seasons and rages like a torrent during the rains. When I did my fieldwork in 2002–2003, the bridge was washed away and negotiations were in place to have it reconstructed, but I was genuinely surprised to learn that this bridge had been destroyed and rebuilt no fewer than five times since Karaurî was elected in 1979. The Thangatha bridge connects Mikinduri with the lowland market centre of Kunatî, itself a major link to Tharaka, Ukambani and the communities living on the fringes of the Meru National Park. The bridge's necessity hardly needs to be demonstrated. It also remains the lifeline for the commercial economy. The area's main agricultural product is tobacco and is grown under concession to British American Tobacco (BAT), who are the single industry and, as such, act as a kind of concessional government, particularly in matters concerning the overall maintenance of infrastructure surrounding the irrigation, processing and transportation of tobacco.

In forty years, the Thangatha bridge had been a thorn in local development, each time it collapsed, eventually being rebuilt in timber, as other reconstructed bridges in the constituency. In Parliament, Karaurî frequently questioned the wisdom of building bridges from timber with such heavy rainfall swelling the size and power of normally placid rivers, as witnessed in 1999 following the damage of *El Niño*:

> I also want to lament that there is a bridge on Road 484 linking Kianjai, Mikinduri, Gatithine and Kunati areas. The Government spent KShs. 4.7 million on that bridge. At that time, I was in Parliament and I was complaining that the materials were being sold and were not used properly but my complaint fell on deaf ears. What happened when the *El Nino* phenomenon came? The bridge was swept away. So, for a long time now we are driving across a big river, on which we have to use the four wheel drive vehicles; that is, either a Land Rover, Land Cruiser, or Range Rover, but then you know that you are also swimming under the bridge and you are ruining it.[1]

My experience of driving across the Thangatha was precisely one of swimming as *The Queen Mother* ploughed through the water while being swept downstream with the current, risking all on a knife's edge, before scrambling up the

1 Kenya National Assembly Official Record (Hansard), 20 July, 1999, Government Printers, Nairobi.

opposite bank with the remaining power of the four cylinder engine, belching steam and complaining from the strain.

D484 would prove to be an enigma to Karaurî and his supporters. The further he pursued engineered concrete bridges and a tarmac road in Parliament, the more expectations of development grew among potential voters, unsettling his popularity and political following. Moreover, the Meru-Mikinduri-Maua road was embroiled in a politics that bled into the twisted issues of land rights and the growth of commercial agricultural, mainly the highly bureaucratised coffee and tea industries, but also the burgeoning local market for cereals, such as beans and maize. In 1976 monies were procured from the tea cess and DDF (District Development Fund) to build access roads high up in the 'tea zones' excised from the forests of the Nyambene massif, but this brought with it conflicts over compensation for agricultural land lost to the building of new road extensions in 1980. Some eighty-odd farmers from Mîciî Mûkûrû had the new tea road, an extension of D484, cut through their land, excising soil and tea stumps alike, but only a handful were ever compensated for their losses. Karaurî rose to champion their cause upon seizing his seat in Parliament, finding loopholes in the Land Acquisition (Amendment) Act and the absence of standardised government claim forms that severely disadvantaged many of his Tigania constituents in trying to get something back for the loss of their property.

Karaurî's Igembe counterpart, Joseph Mwenda Malebe, found that his constituents were in an even worse conundrum, since land adjudication and the issuing of title deeds had not been effected throughout much of Igembe by 1990. This resulted in a situation where the building of tea roads along the Mikinduri-Akachiû-Maua segment resulted in compensation that was 'peanuts because you do not have a title deed' and where those land owners who could corrupt officials with their documentations became the 'tycoons in the area'.[2]

Like D484, Karaurî's political career lay in ruins by the time I lived in Mikinduri in 2002 and the December General Elections would see him lose his parliamentary seat to a virtually unknown Mûthaara youth, Peter Gatirau Munya. One of my clearest memories of those early days was stopping *The Queen Mother* to ask people walking on the road, carrying heavy loads, if they wanted a ride. It proved an effective way of making friends, to be sure, but nobody seemed to miss an opportunity to pass vitriol onto Karaurî and KANU through the ruins of this backwaters road. Public transport fares were prohibitive for many people owing to the outlay in spare parts most vehicle operators had to

2 Kenya National Assembly Official Record (Hansard), 17 April, 1990, Government Printers, Nairobi.

fork out. Yet, in my reckoning, by the time the violent clashes of the December 2002 elections had passed into memory (there was no violence in Meru Town) and people stopped talking about Mikinduri and Mûthaara rivalries (if only for a while), the burning political issue that would remain in Tigania, at least until my return nearer the end of the decade (2007 and 2008), remained the main trunk road and its unclassified access routes.

While this political history is a necessary background to the untold story of a tiny rural road of little national importance, to the people of Tigania East, D484 was the materialisation of their neglect at the hands of the government and, sometimes, their representative leaders. I can maintain however little D484 figures in politics at a national level, rural roads like it far outnumber the paved road network in Kenya. It is in this descriptive sense, then, that stories like that of D484 are important to tell in any discussion of this East African country's infrastructural politics. In what follows, I turn to an altogether more ethnographic consideration of being 'on the road', presenting brief vignettes of informal bridge construction and road repair as illustrations of infrastructural sociality.

Being Social the Infrastructural Way

With the end of the rains in March 2002, getting in and out of Mikinduri was a serious challenge to drivers and passengers. The loss of two bridges along D484, whose foundations were swept away by high waters, meant that only two remaining routes were still passable if one wished to reach the main tarmac road in the county linking Meru Town to Maua: the tea road that passed over the top of Mîciî Mûkûrû, high up on the Nyambene massif; or, the winding and heavily damaged, but much shorter route, through Mlathene onto Kianjai, one of the main markets located along the paved road in Tigania West.

During one of the heaviest weeks of rain, *The Queen Mother* had become bogged down trying to reach Maua along a ruined section of the Mîciî Mûkûrû tea road and 'burned out' the clutch plate. Needing this and other spare parts to repair her, I was forced to travel to Meru Town by public transport. By public transport, I mean the three 24-passenger Isuzu buses, roughly six old Land Rovers, and a dozen or so lorries that operate around vague schedules that depend largely on the weather and the amount of people or goods to be transported. At this point in the rainy season, however, only the Land Rovers could be relied upon to get through to the tarmac road.

'Mwalimu' ran one of these Land Rover services from Mikinduri to Kianjai, where he lived and operated an agricultural input and hardware shop. Drivers like Mwalimu will not travel until their vehicles are crammed with people and the roofs loaded high with bulging sacks of beans, dumbfounded and bleating goats and clusters of bound-up chickens. This initial step in travel is called *Amua, Twende*, 'Agree, Let's Go' (like the PSV described above) owing to the competition between different vehicles touting for passengers and the frustrating practice of travellers to suddenly rush out of one vehicle to board another thinking it to leave earlier. With me crushed with another passenger and a small school boy into the front passenger seat and at least eight other passengers in the back, we were still being loaded on the roof as Mwalimu began to move.

After negotiating the slick muddy incline out of Mikinduri, Mwalimu drove through the forest to a notorious hairpin turn at a particularly treacherous place where the culverts lay disarticulated under a slop of liquefied mud. Taking his chances, Mwalimu drove gingerly into the turn in low gear, but suddenly pitched over violently when the front driver's side became lodged against a broken culvert. Mwalimu immediately tried to reverse aggressively before the vehicle sank deeper, the whine of the gear box and the wheels gurgling under mud sharply punctuated by the distinctive snapping sound of aluminum.

During the rains, the roads connecting Mikinduri to the rest of the world undergo what can be described as a 'cascading failure', comparable to major disruptions in urban infrastructure like blackouts, traffic jams and leaking sewerage. The notion of a cascading failure is that one disruption moves through the whole system (Graham 2010: 18). Consider Stephen Graham's discussion of cities' reliance on electricity:

> In an electrical blackout it is not just electric lighting that fails. Electrically powered water and sewerage systems tend to grind to a halt. Public transportation stops. Food processing and distribution is disabled. Health care becomes almost impossible. Even the Internet ceases to function.
>
> GRAHAM 2010: 18

While ethnographers have paid considerable attention to the infrastructural 'provisionality and uncertainty' in African cities (Larkin 2008: 181; Simone 2004), very little attention has been afforded to the vast rural hinterlands where the majority of the population lives and through which major revenues are generated and extracted where passable road and effective communications infrastructure is in place and in technical order.

Missing from the larger discussion about 'cascading failures' of electrical, waste, telecommunications and transport systems is the effect these have on the transformation of 'ordinary' sociality. In their considerations of 'what happens' when infrastructures fail, urban ethnographers frequently focus on the functionality of infrastructure and, as such, they skip over how infrastructural disruptions are productive of shifts within everyday sociality.

Returning to Mwalimu's stuck Land Rover, and by way of furthering the notion of infrastructural sociality touched upon above, I want to suggest here that the rural residents of Tigania suffer cascading failure as a *normal* condition of their social life.[3] We can look at this normality in several ways. In the most obvious sense, Mikinduri's poor road infrastructure affects even the metabolism of everyday life, in the very functional sense of transporting the basic commodities, like food and medicine, to those living in this area. During my fieldwork, when the rains came, vulnerable people such as expecting mothers, children and infants died of acute illnesses because there was no way to transport them to hospital for medical treatment. In overall terms of economic development, the roads support or maintain the agricultural or commercial activities that generate income and their ruination is experienced as rotting foodstuffs, expensive transport costs and untimely delays of many necessary transactions. But this normality of failure and disruption is too easily seen in purely functional terms, begging an alternate set of interpretations. The scales of (dis)connection do not simply radiate outward, linking small rural Tigania to the complex global economy and its wider political-historical horizon; it also comes to condition experience and thereby transform who the Tigania historically consider themselves to be: that is a people who have been neglected by government and exploited by their political representatives. For the small crowd of people milling around Mwalimu's trapped vehicle, this disruption was the product of political tensions going back to colonial times; but it was in the *modus vivendi* of their everyday lives that disruption was seen as an integral, banal and necessary part of the journey. This was expressed by Mwalimu himself who expected that each person present knew, more or less, that reaching their tarmac destination meant that they would have to somehow assist in bringing the mired Land Rover out of the broken culvert and back onto the road.

Several young men emerged out of the forest, carrying a chainsaw and a small jerry can, presumably charcoal burners. They struggled to walk along higher, relatively drier parts of the road until they stood speaking with Mwalimu, still behind the wheel of the now-idling but quasi-submerged Land Rover. A man standing next to me expressed his apprehension about their arrival, since roadside brigandry was commonplace along this section of the road through the

3 I am adapting these ideas from 'normal accident theory', which pervades the literature on the 'accidental' and defines, in part, the iatrogenic logic of technological development and innovation (Perrow 1999).

forest, robberies carried out by felling tree trunks in front and behind passing motor vehicles. Within a short moment, however, the young men were asking us to help them as they carefully assessed how to get the four-wheel drive out of its predicament. They would receive the equivalent of a day's wage for this if they were successful.[4] With the three women, the school boy and an elderly man tacitly excused from the task, we climbed into the turn, taking caution not to fall down into the broken concrete under the mud and relayed information from Mwalimu and the young men when to push. Fifteen minutes of arduous (and dangerous) work proved only to unsettle the Land Rover and it leaned even more precariously on its side.

At the height of our discussion, which combined what else we could do with a series of complaints about government neglect, we heard another vehicle approach from the direction of Miathene, an unloaded Canter lorry owned by a prominent Coca-Cola and Kenya Breweries distributor. The driver spoke very briefly with Mwalimu, then with great strain manually withdrew the Land Rover's winch cable, hooked it onto the chassis of the lorry, and once back behind the wheel, reversed until the cable was taut. With a rocking motion that was both hopeful and precarious, both drivers worked their vehicles until, very suddenly, Mwalimu drove out of the deep hole.

I have gone into this one episode in grainy detail to demonstrate what was quite a normal journey during the rains in this part of Meru. Unlike Tim Ingold's (2007) two distinctions in theorising 'modalities of travel', wayfaring and transport, in which he describes the latter as 'destination-oriented', passengers and drivers along this and other local motor roads engage with infrastructural failures in ways that resonate or touch upon Ingold's alternate notion of wayfaring, specifically with respect to the interruptions of movements, rest, uncertainties and displaced temporality that characterised transport in Tigania. Travel along these roads is to experience tension.

Ingold's distinction echoes an older one made by Michel de Certeau (1984) who spoke of the experience of a railway passenger as a 'traveling incarceration' where the 'unchanging traveler is pigeonholed, numbered, and regulated' (de Certeau 1984: 111). It is not my intention to gainsay these valuable discussions, but neither can my ethnography of life in Tigania accord with their conclusions that transport is either completely destination-oriented or a form of incarceration. There certainly are instances where this indeed is the case. During

4 In 2002, a day's wage (*kîbaruga*, or Swahili, *kibarua*) was a 100 KSh, roughly $1.30 USD, roughly half of what it cost a single passenger as a fare from Mikinduri to Kianjai. On a fully loaded Land Rover, drivers might expect to receive 3,500 KSh on a single trip, about 15,000 KSh for two return trips on a good day, such as during the dry seasons.

the time of my fieldwork, travelling between towns or cities along the tarmac in overloaded, speeding and poorly maintained *matatu* did feel a bit like being imprisoned, indeed, like a breath-taking stint on death-row. But for the people of Mikinduri, transport was not so much about getting to one's destination in any facile sense, because one was cast into a situation where, as passenger, one had to participate directly in making it to that destination rather than being passively transported there. In other words, and to parse those of Ingold, travelling by a PSV Land Rover in Tigania was to dwell in a sustained moment of tension. It was 'like holding one's breath' (Ingold 2007: 77), where there is no fixed point-to-point connectivity between places because sometimes one has to go off the line of the road to overcome some obstacle, where there is no sense of being 'temporarily exiled' because one is a direct participant in the journeying, where the dangers along the road, real and perceived, mean that no series of precautions are sufficient enough, sustaining the sense of tension and its productive, if forced infrastructural sociality.

There is also another sense in which road transport is lightly dismissed by anthropologists as being of little interest or insight. This is because infrastructure and its materiality lie largely invisible when, as a system, it functions as it should. Take this away, as in instances when people first must repair or maintain infrastructure, this materiality takes on different meanings and provokes all kinds of tensions that expose the centrality of roads and related infrastructure in the way that we live. For the people of Tigania, these access roads were tied in with their communal value of self-achievement through development. At the time I undertook fieldwork in Tigania, many people expressed to me how they had given up hoping that government would bring development and sought ways to strike a 'good-enough' *modus vivendi* with the ruined roads, culverts, drifts and bridges through forms of action that were independent of government officials or politicians. Infrastructure and its tensions, in this case, the ruination of Tigania's road 'network', seemed to galvanise the people of Tigania into a political community prior to the eventual arrival of a tarmac road in 2009, with infrastructure remaining the key community issue to date. In travel along the roads, this political community was most keenly felt, as when, people made stringent efforts to find ways of resolving permanent or temporary disruptions, often by repairing or maintaining what was damaged. This coalescent idea of community, however, was not without key political tensions or factions that make the repair and maintenance of road and transport infrastructure a focus of conflict.

In the following section, I spare more detail to this idea of political community and conflict through examining the repair of a bridge by an organised group of young men and their conflict with representatives of the state and local businessmen.

Building Bridges

I called it the 'piano bridge' to amuse my passengers whenever I crossed the wooden bridge across the Thanantû river. As any vehicle proceeded across its timber surface, the roughly hewn poles would move up and down like the keys of a piano. With an unladen weight of about 1,300 kilograms, *The Queen Mother* bobbed left and right crossing this bridge, a sensation not unlike being on board a small sailboat on choppy waters. Constructed entirely from the softwood gravellia, an exotic tree grown throughout Kenya, this wobbly little bridge now took on the weight of all of Mikinduri's through traffic, some hundred vehicles per day.

The bridge was built by a group of young men who lived nearby, several of them working in the local transport industry. After coming back from a few days visiting friends in Meru Town, I found that the old concrete bridge had come away from the river's bank and lay half sunk into the silted waters of the Thanantû, one of the larger rivers near Mikinduri. No vehicle can cross the Thanantû due to its steep and sandy banks, constantly shifting with the rise and fall of the river. On this day, I turned *The Queen Mother* back towards Meru Town, eventually finding an alternative route back out to the tarmac at Kianjai, through the tea estates at Mîciî Mûkûrû, then onto Mikinduri: a route of over 55 kilometres, instead of the five remaining kilometres to town from the bridge itself. The young local men saw a seasonal opportunity with every rains, reconstructing temporary bridges at various points along D484. And, as usual, when they set to construct it, cutting down their own trees and sourcing their own labour, they were under no illusions as to what they were constructing: a toll bridge.

Initially, drivers applauded the youths' efforts and initiative, even accepting the toll of 50 KSh as preferable to the extended and profit-losing routes over the Nyambene massif. Operators of the larger Isuzu passenger-buses had lost a great deal of their profits for the two weeks it took to construct the bridge at Thanantû and one of them, the bursar for a local girl's secondary school, told me, 'I'm grateful for these boys, they're doing what our MP should be doing.' Having paid several tolls to the young men that manned the bridge, carrying out repairs and guiding heavier vehicles across, I was eventually told to pay a 'season ticket', a one-off payment, I thought, of 1,000 KSh.[5] Locals I spoke with about the bridge and its more-or-less illicit toll eventually began to complain that they had paid for a season ticket, but then were being charged 'extras'

5 In 2002, this would have been about $15 USD, but was considered an expensive charge in Mikinduri at the time.

depending on what they were carrying. The group's organiser, a young man who used to be a *matatu* tout for one of the Isuzu buses, kept a register of all those drivers who had paid the lump sum, but he was not always present at the bridge, leaving the other youth to increasingly exploit travellers and transporters. In response, some PSV operators increased their own fares, leading to a range of complaints from passengers themselves. .

The first open conflict over this toll bridge came from government officials, particularly the District Officer (DO) and the Officer Commanding Police Station (OCS). As I heard about it, both the DO and OCS were returning to Mikinduri in a private car and upon reaching the bridge refused to pay the toll because they were on 'government business'. They had, in fact, been drinking at a small pub in Kagaene and in the confrontation with the youth they threatened to arrest them unless they were let to pass. The second conflict started when local businessmen met with the group's organiser to discuss why 'season tickets' were no longer being honoured and raised concern that extortion was being reported to them. Feeling pressured by these older men, the youth argued that they had used their own trees to further repair the damage caused to the bridge by the businessmen's heavy lorries and buses, claiming that any extra costs were legitimate for the maintenance of the bridge. Piqued by the adamant stance of the businessmen, the youth's leader threatened to tear down the bridge completely, once again cutting Mikinduri off from the rest of the county.

These conflicts may seem to centre on the legality of the toll itself and the charges of extortion, but they have more complex social roots. With respect to the DO and OCS, there is a deep historical antagonism between local people and government employees, especially administrators and security officers that stretches back to colonial times. In particular, the police forces are from time to time embroiled in various sexual scandals and, on occasion, accused of violent assault or even murder of people in their custody. The DO at the time was particularly hostile to local people, commandeering their vehicles (and mine), persecuting youth during the circumcision and initiation period and running up huge bar tabs he refused to pay.

While the conflict at the Thanantû bridge between the youth who built and maintained it and the DO and OCS can be explained within a larger political context, that of Kenyan ethnic ideology and the violence of the provincial administration, their confrontation with the local businessmen demands understanding something about Meru age-set formation and the gerontocratic ideology that pervades everyday familial and clan politics. For the youth at the bridge, the compulsions weighed against them by their elders were extremely complex, fitting into an ethos and ethics articulated on the force of elders' 'curses' or 'blessings' (*kîruume / kîrathimo*). The businessmen were attempting to assert their agency over that of these local youth, trying to reclaim the bridge as if were part of the 'commons' and a public good. On this occasion, these youths were particularly refractory, threatening to destroy what they created, but they eventually acquiesced their 'ownership' over the bridge and it came to be maintained by casual day labourers hired, in the main, by the commercial users of the bridge. This resulted in a compromise, as the day labourers turned out to be a handful of the same youth and the road users the same businessmen.

This question of ownership over infrastructure raises further investigation into the nature of governance in Tigania and, indeed, throughout the country. In introducing the political history of D484, where the bridge over the Thangatha river became a source of conflict on the route to the tobacco-growing lowlands, it was asserted that BAT exercised a kind of concessional power over this stretch of road and its supporting infrastructure. Achille Mbembe's (2001) notion of 'private indirect government' is pertinent here. My intention is not to support his larger claims, but to focus most explicitly on his claim that – sometime in the 1980s – African states lost their regulatory capacities that shored up their legitimacy in the eyes of the citizenry:

> Having no more rights to give out or to honour, and little left to distribute, the state no longer has credit with the people. All it has left is control over the forces of coercion, in a context marked by material devastation, disorganisation of credit and production circuits, and an abrupt collapse of notions of public good, general utility, and law and order.
>
> MBEMBE 2001: 76

Cemented into Mbembe's thesis on private indirect government is the privatisation of sovereignty. As explained above, Tigania East's long-standing MP, Karaurî's own parliamentary saga with the bridge at Thangatha represented the loss of people's faith in government promises to bring development. By the time I was driving through this landscape, a new source of exclusively infrastructural development was emerging from private companies, in this case, the multinational giant BAT. Similar processes of transformation in the ownership and privatisation of services was occurring across the infrastructural spectrum from telecommunications to energy provision.

On the road to Kunatî in the lowlands where people grow and process tobacco under consignment for BAT, I saw engineers and local labourers repairing the rough road, slowly hauling in truckloads of gravel and grading sections that had not seen any maintenance in its three decades of existence. Furthermore and much to my surprise, the washed out bridge was being reconstructed in concrete by a private construction company wholly financed by BAT. Although officially registered through the Roads 2000 Programme, one of the local foreman I spoke with told me that their wages were paid through a BAT employee. The bridge was an impressive project to watch develop. While for the first several weeks, vehicles continued to drive through the Thangatha river itself to reach the Kunatî side, BAT lorries consistently delivered building materials, the company managers erected a small kiosk at the site, and labour recruiters hired dozens of local men to assist in the building of the bridge with startling efficiency.

The Republic of Kenya's sovereignty during the period of time I lived in Tigania consisted of a heavy-handed police presence, laws and litigation, a modicum of schools originally opened by mission societies, the battered road, some bridges, a few water cisterns and a mobile-phone mast far in the distance. In reference to Michael Mann's (1986) notion of the state's 'infrastructural power', vested in the power to exercise sovereignty through a continuing expanse of material and administrative reforms, the Government of Kenya (GK) during the late 1990s and early 2000s seemed like a ghost state with reference to road repair and maintenance. In the absence of government maintenance, we can see the materiality of state effects in ruination (see Harvey 2005). Along the extension of D484 to the bridge at Thangatha, it was the absence of the state's presence that was most felt, with the construction of a newly repaired road and construction of a permanent bridge a project 'owned' and 'operated' by BAT and seen by local people as an alternative to the state. I left Tigania before the bridge was completed, but I have heard that the market in Kunatî has grown, with shops and kiosks mushrooming with the completion of the bridge and an all-weather gravel road reaching into Ukambani. There are now PSVs taking passengers to Mwingi and, from there, onto Garissa, opening up a fast, secure transport and trade corridor to the rest of Eastern Kenya. This eastern extension of D484 had been one of the original development plans for postcolonial Kenya in 1964, but it is only in 2012 when local residents are capable of travelling or transporting their goods in this direction.

Conclusions

The period in which my fieldwork in Kenya was undertaken offered a glimpse of a crisis in state sovereignty. In this chapter, I provided the examples of how fare-paying passengers participated in getting bogged down Land Rovers out of the mud and back on the road, how local youth seized opportunity to build an *ad hoc* toll bridge, or how a major multinational corporation effectively took over the construction and maintenance of significant roads and bridges from state government authorities. Through these examples, I have aimed to afford the notion of infrastructural sociality some ethnographic contexts. This concept is aimed at opening up opportunity for looking at infrastructure in relational terms, looking at how disruption and ruination compel new social relations to emerge in moments that are interstitial, often fleeting and unstable. Anthropological considerations of infrastructure point to the temporality of infrastructure's seeming obduracy, showing through historical and ethnographic considerations the instability underpinning most fixed infrastructure. It is not only that breakdown in everyday services makes infrastructure visible, for examples, in the supply of electricity, commodities, waste flows, water or fuels. But the conditions provoked by such breakdown compel, sometimes for very brief moments, surprising new kinds of sociality and forms of relationships that often temporarily suspend and invert expectations between people and things. When in highly competitive economies we witness drivers co-operate together to assist one another to get out of a ditch, or travellers and local residents work together to clear electrical cables from blocking a road, we need to rethink some of our assumptions about what happens when normal accidents or infrastructural breakdown occurs.

On the other hand, this chapter has aimed at demonstrating how roads, in particular, are built according to a cluster of imagined economic purposes and political promises, only to be negated by the dialectics of the conditions the building of the road produces, as it becomes part of the wider political-economy. Writing about the incompleteness of road building in the Amazon, Jeremy Campbell insightfully informs us how a road has the 'singular ability to replace the conditions of its creation' (Campbell 2012: 496). This is an insight because sometimes roads and associated infrastructure lead, unexpectedly, to the undermining or destruction of the purposes for which they were built. Across rural Kenya, for example, the telephone poles that line the edge of roads, with their sagging wires, were never really put into widespread usage, before cellular and mobile technologies made landlines redundant and replaced them. But the analysis of this technological and infrastructural negation of fixed-line telecommunications is incomplete with examining how demand for access to telephones led to a kind of overload in what the national telecommunications network could carry. The building of the nation's rural roads was also similarly linked to an imagined agrarian transition in the 1970's that could never have predicted that the massive fiscal borrowing that fuelled the coffee and tea industries' early successes would be undermined by the processes of paying back run away debt under the gavel of structural adjustment and neoliberal reform. The economy of disappearances I've described relates to the affective haunting of the Meru countryside by a historical vision of happy coffee and tea farmers, educating their children, attending church services and government *baraza*, marketing their produce through co-operatives, paying their cess and tax, while travelling by private pick-ups along regularly maintained and graded bitumen-quality roads. This imagined landscape of what Jeremy Campbell (2012) called 'smallholder utopia' came with an infrastructural wave of changes in Tigania during the 1970s and 1980s, referred to as 'agrarian transition' and meant to be a permanent condition (Berry 1993), provided the political-economic context for its own infrastructural decay.

When I first visited Meru in 1991, the Kenya described by Angelique Haugerud (1997) – of a 'culture of politics' – presented a *figural* of a very fragile political and economic order, shrouded by the very dramatic resurgence of multi-partyism and the ideological returns of democracy and civil society, especially into rural areas of the country. By 1998, when I first carried out field research on the economic liberalisation of coffee farming and examined the disintegration of co-operative marketing societies, it was apparent that this developmental regime was slated for radical, highly disruptive change and that Berry's views on the dissolution of agrarian transition were, in part, borne out by empirical facts. It was also the period of time that many anthropologists were writing on globalisation, modernity and, latterly, neo-liberalism, terms that would take on polyvalent registers as the decades wore on. I suspect part of what I was researching among coffee farmers in the late 1990s was the figural of a vaguely defined, newly emergent kind of economy, but one to which no development plan could pinpoint. It was likely going to favour individual entrepreneurs and corporate investors, as well as be concentrated largely in urban parts of the country, but it was also bolstered by far-reaching transformations in both road infrastructure and telecommunications.

Examining the wretched state of Meru's 'tea roads' during my fieldwork of 2001–2003, I now make the argument that what I was viscerally witnessing through travel and journeying through this agrarian landscape was the ruins of a 'no-place', a smallholder utopia, in which actors such as Karaurî played a major role in fostering and then defending when it came to its demise. I am thus moved intellectually by Jeremy Campbell's (2012: 496) insightful analysis of an Amazonian highway and the fragile materiality of most infrastructure when not routinely maintained, seeing along D484 how a road can bring about circumstances of use or consequence that obviate the initial purposes for its construction. While I do not know what will replace the economy of disappearances with the paving of D484, a minor road in the scheme of things, I assume that it will be built upon the ruins of what was once a deeply inculcated dream among local farmers, as real as the various bridges built across the Thanantû river, but sharing in these materialities also the speedy shift from obduracy to fragility that accompanies radical economic transition and its political appearances.

References

Berry, Sara. 1993. *No Condition is Permanent: The Social Dynamics of Agrarian Change in Sub-Saharan Africa*. Madison: University of Wisconsin Press.

Campbell, Jeremy M. 2012. 'Between the Material and the Figural Road: The Incompleteness of Colonial Geographies in Amazonia.' *Mobilities* 7 (4): 481–500.

Dalakoglou, Dimitris and Harvey Penny. 2012. 'Roads and Anthropology: Ethnographic Perspectives on Space, Time and (Im)mobility.' *Mobilities* 7 (4): 459–465.

De Certeau, Michel. 1984. *The Practice of Everyday Life*. Berkeley: University of California Press.

Ferguson, James. 1999. *Expectations of Modernity: Myths and Meanings of Urban Life on the Zambian Copperbelt*. Berkeley: University of California Press.

Fontein, Joost. 2011. 'Graves, Ruins, and Belonging: Towards an Anthropology of Proximity.' *Journal of the Royal Anthropological Institute* 17 (4): 706–727.

Gluckman, Max. 1940. 'Analysis of a Social Situation in Modern Zululand.' *Bantu Studies* 14 (1): 1–30.

Gordillo, Gastón R. 2014. *Rubble: The Afterlife of Destruction*. Durham: Duke University Press.

Graham, Stephen (ed). 2010. *Disrupted Cities: When Infrastructure Fails*. New York: Routledge.

Graham, Stephen and Thrift Nigel. 2007. 'Out of Order: Understanding Repair and Maintenance.' *Theory, Culture & Society* 24 (3): 1–25.

Harvey, Penny. 2005. 'The Materiality of State-Effects: An Ethnography of a Road in the Peruvian Andes.' In *State Formation: Anthropological Explorations*, edited by Krohn-Hansen Christian and Nustad Knut, 123–141. London: Pluto Press.

Harvey, Penny and Knox Hannah. 2012. 'The Enchantments of Infrastructure.' *Mobilities* 7 (4): 521–536.

Haugerud, Angelique. 1997. *The Culture of Politics in Modern Kenya*. Cambridge: Cambridge University Press.

Hetherington, Kregg. 2014. 'Waiting for the Surveyor: Development Promises and the Temporality of Infrastructure.' *The Journal of Latin American and Caribbean Anthropology* 19 (2): 195–211.

Humphrey, Caroline. 2005. 'Ideology in Infrastructure: Architecture and Soviet Imagination.' *Journal of the Royal Anthropological Institute* 11 (1): 39–58.

Ingold, Tim. 2000. *The Perception of the Environment: Essays on Livelihood, Dwelling and Skill*. London: Routledge.

Ingold, Tim. 2007. 'Earth, Sky, Wind, and Weather.' *Journal of the Royal Anthropological Institute* 13 (1): 19–38.

Kernaghan, Richard. 2012. 'Furrows and Walls, or the Legal Topography of a Frontier Road in Peru.' *Mobilities* 7 (4): 501–520.

Lamont, Mark. 1999. *Not yet Soko Huru: The Local Appropriation of Free Market Discourse in the Coffee Industry of Rural Kenya, Meru District (1998)*. MA thesis, Concordia University.

Lamont, Mark. 2012. 'Accidents Have no Cure! Road Death as Industrial Catastrophe in Eastern Africa.' *African Studies* 71 (2): 174–194.

Larkin, Brian. 2008. *Signal and Noise: Media, Infrastructure, and Urban Culture in Nigeria*. Durham: Duke University Press.

Mann, Michael. 1986. *The Sources of Social Power, Vol. 1: A History of Power from the Beginning to AD 1760*. Cambridge: Cambridge University Press.

Mbembe, Achille. 2001. *On the Postcolony*. Berkeley: University of California Press.

Miyazaki, Hirokazu. 2004. *The Method of Hope: Anthropology, Philosophy, and Fijian Knowledge*. Stanford: Stanford University Press.

Navaro-Yashin, Yael. 2009. 'Affective Spaces, Melancholic Objects: Ruination and the Production of Anthropological Knowledge.' *Journal of the Royal Anthropological Institute* 15 (1): 1–18.

Perrow, Charles. 1999. *Normal Accidents: Living with High-Risk Technologies*. Princeton: Princeton University Press.

Roseman, Sharon R. 1996. 'How we Built the Road: The Politics of Memory in Rural Galicia.' *American Ethnologist* 23 (4): 836–860.

Simone, AbdouMaliq. 2004. 'People as Infrastructure: Intersecting Fragments in Johannesburg.' *Public Culture* 16 (3): 407–429.

Street, Alice. 2012. 'Affective Infrastructure: Hospital Landscapes of Hope and Failure.' *Space and Culture* 15 (1): 44–56.

Tsing, Anna L. 2005. *Friction: An Ethnography of Global Connection*. Princeton: Princeton University Press.

Tsing, Anna L. 2015. *The Mushroom at the End of the World: On the Possibility of Life in Capitalist Ruins*. Princeton: Princeton University Press.

Verbeek, Peter-Paul. 2004. *What Things Do: Philosophical Reflections on Technology, Agency and Design*. University Park: Pennsylvania State University Press.

Wasike, Wilson S.K. 2001. *Road Infrastructure Policies in Kenya: Historical Trends and Current Challenges.* Kenya Institute for Public Policy Research and Analysis.

CHAPTER 8

Negotiating Desert Routes

Travelling Practices on the Forty Days Road

Rami Wadelnour[1]

Introduction

Africa's desert roads often lack physical demarcations. Drivers on these unpaved roads pick their own ways and follow tracks that sometimes divert far from others. Desert roads may thus be several hundred meters and even kilometres wide, and roadsides are typically fluid as roads peter out into the open desert. This fluidity is reflected in the travel practices that are performed on and along the roads. This chapter attempts to examine the processes of *road-making* on one of these desert hinterland roads, namely on the 'Forty Days Road' in the Sudan. I will argue that this road is in a very literal sense *made* by its users. The road itself is a result of travel practices. It is produced by the traffic that passes on it, and only the road that is travelled on persists. The road that is abandoned is soon obliterated by nature. Complementary to the other chapters in this volume that address the orders of the road (see in particular Bize, Bürge, Klaeger and Stasik), in this contribution I look at road-making by attending to the materiality of the road itself.

I further argue that African desert travel is principally a form of collective travel and, more so, collectively organised travel. Moreover, transport is not differentiated, in the sense that goods as well as passengers are transported on the same trucks. The chapter thus offers an ethnographic account of a case of collective long-distance travel in Africa. The 'fluidity' of the road, combined with the collective and undifferentiated nature of desert travel, provides for a unique experience of those using, and hence producing, it. By focusing on the experiences of travellers between two cities in the Sudan, this chapter aims at providing an ethnographically grounded analysis of travel practices on a long-distance desert road.

1 The present findings are based on eleven months of fieldwork, conducted in 2011 and 2012 as part of the project 'Roadside and Travel Communities' within the framework of the SPP 1448 'Adaptation and Creativity in Africa' funded by the Deutsche Forschungsgemeinschaft. I would like to express my gratitude to Kurt Beck for his support and guidance. I would also like to thank Michael Stasik and Gabriel Klaeger for their valuable comments.

 | DOI 10.1163/9789004339040_009

The Forty Days Road

Today's Forty Days Road received its name from the historic Forty Days Road, in Arabic *Darb al-Arbaʿîn*, which was named according to the time it took to travel from Darfur to southern Egypt. While resembling its ancient predecessor in various regards, today's road follows different paths. In contemporary language usage, the Forty Days Road is also referred to as *al-khatt* (Arabic: line) and as *as-sikka* (Ar. track). As the map below clearly shows, this desert road connects Omdurman with El Fasher in Northern Darfur, while its many branches pass through small towns in the state of Northern Kordofan, such as Soderi, Hamrat ash-Shaykh and Umm Badir, as well as various smaller villages. The approximate length of the road is 800 kilometres. According to my own travel experiences, the full journey can take from four to seven days. As illustrated on the map, to the south of the road, an asphalted highway connects the same destinations, Omdurman and El Fasher. Passing through the towns of Kosti and El Obeid, the paved road ends in En Nahud where an unpaved road continues to El Fasher.[2] The paved part of this road allows travellers to cross the distance between En Nahud and Khartoum in ten hours. In spite of the considerably longer time spent on the Forty Days Road, its very fluidity provides for its users advantages which counterbalance the inconveniences of the long – and certainly arduous – journey.

Travelling on the Forty Days Road is characterised by a number of features that are distinct to desert travels, which include the traversing of vast distances on unsurfaced tracks across sparsely populated areas; a non-differentiated transport of passengers and goods, wearing climatic conditions, as well as high unpredictability in terms of security and heightened dangers related to mechanical breakdown. Institutions along the road such as convoys, checkpoints and truck stops, serve as means for reducing related uncertainties. The state is scarcely present in these areas, which veritably qualify as 'hinterlands'. Here, it is the road users who conquer the space, form institutions and control conduct in ways and modes that remain largely outside of state regulation (see Beck 2013).

Similar to other hinterland roads in the Sudan, the majority of vehicles present on the Forty Days Road are trucks modified in order to serve the specificity of the journey and the roads. One of the main functions of these modifications is to enhance the capacity of the vehicles to carry more freight. Most of these lorries were originally designed as dumper trucks with a manoeuvrable open-box bed. In order to increase their loading capacity, local blacksmiths

2 In March 2015, the paved road to El Fasher was finally completed.

and mechanics remove specific parts, such as the box bed, and extend and reinforce the chassis (Beck 2009; Hänsch 2009).

Experiencing the Road

Travellers on the Forty Days Road spend days traversing the desert while occupying limited spaces on the back of a lorry and clinging to its load. Sitting on the top of the lorry, they are exposed to the heat and dust of the desert. When exhausted and falling asleep, they risk falling from the moving vehicle. Only during rest stops are travellers able to get down from the vehicle to have their meals, pray, rest and prepare for the next part of the journey. During the journey, their daily rhythm and bodily functions, and indeed their overall existence, are forcefully aligned to the rhythms and functions of the lorry. Besides the apparent physical discomfort, travellers are generally not aware of the exact places they are travelling through, which thus adds an element of psychological distress to the journey. Such is the uncertainty of desert travel that even

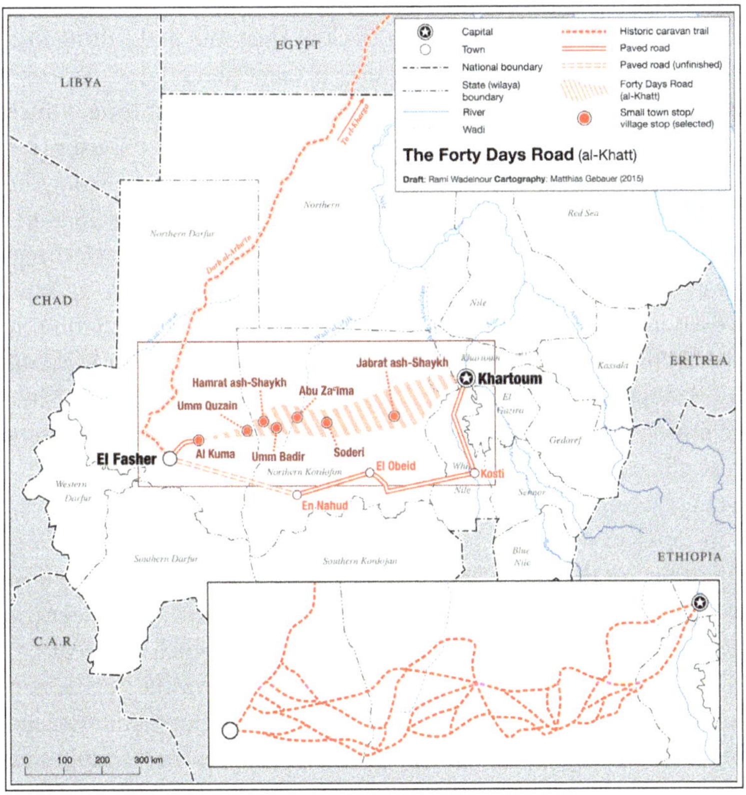

MAP 8.1 *Area covered by the Forty Days Road*

drivers at times hold a guess rather than a clear idea about the time of arrival at the next stop on the journey. Air and sea travellers also share such uncertainty of travel since the nature of the environment and its 'fluid' surroundings do not allow deciphering landmarks that would provide clues as to the whereabouts in the travel situations (see Ashmore 2013). However, in the case of most air and sea travels today, constant updates about coordinates and the remaining time to destination are offered in order to minimise the traveller's distress. On desert roads, by contrast, such coordinates are unavailable both for the driver and for the passengers. Notwithstanding these differences, the discussion in my chapter is based on the premise that long-distance desert travel establishes a setting similar to the experiences of expeditions and ship crews at sea (Gerstenberger and Welke 2004; Hutchins 1995; Kleinert 2009). Like ocean travellers, desert travellers develop a relationship with their truck, which is close to a home in a vast solitude. They tend to form a moving social space, which could be imagined as a 'community of fate' (Goffman 1986; Klaeger 2009).

Security plays a further, and crucial, role in shaping the travel experiences. While traversing scarcely populated areas without exact knowledge of their whereabouts, travellers on the Forty Days Road may have to face mechanical defects, as well as the danger of losing track of the route and getting lost. Furthermore, they are also exposed to the ever-present potential of extortion and ambushes. In terms of security, the road offers two different zones which are delineated by the borders between the states of Kordofan and Darfur (see on Map 8.1 the area between Umm Quzain and El Fasher). Upon reaching the borders of Darfur, the town of Umm Quzain, security constraints begin to define the whole travel experience. Since the escalation of hostilities in Darfur in 2003, the western segments of the road – mainly within Darfur – are normally only travelled as part of an escorted convoy. These convoys are commonly formed by local pro-government tribal militias known as border guards, *'ḥaras al-hidûd'* (Beck 2013), and after 2013 as 'Rapid Reinforcement Forces'. While travelling in escorted convoys imposes a stricter regime upon the travel community, it also adds to the general sense of insecurity and danger.

Long-distance desert travel is largely composed of practices, which travellers devise in order to make a safe and successful journey. The 800 kilometres of the journey on the Forty Days Road are navigated by one driver who is normally supported by one or two assistants. The assistants never steer the vehicle. They serve the driver in an auxiliary way through which they learn to become drivers themselves. The driver is solely responsible for the vehicle and the load he is transporting. His primary task is to deliver the lorry and the load safely and timely (although the actual time of delivery is rather negotiable). While the driver is the one who is ultimately responsible, it is the whole of the

FIGURE 8.1 *Loaded lorry and travellers resting at an isolated truck stop*
SOURCE: R. WADELNOUR, 2013

crew as well as the passengers that together participate in the *making* of the journey. The kind of 'fated' community renders the desert journey a collective endeavour in more than one sense: passengers are not only transported collectively, they also take an active part in the collective organisation of the journey.

In the remainder of this chapter, I will analyse this collective experience by focusing on three distinguishable bundles of travel practices. The first bundle comprises practices related to the *loading* of the vehicle, which primarily includes the placing and valuing of transported items as well as the seating of the passengers. As I will show, loading is constitutive for the handling of the lorry on the road and for the safety of the freight in several respects. The second bundle consists of practices that I frame by the notion of *passengering*. Principally, these include the practices of 'making' and negotiating the social space of the lorry. The description of the practices of passengering further demonstrates some of the central experiential aspects of the travel, in particular the co-production of the journey by the passengers. The third bundle comprises practices of *driving*, which mainly pertain to the navigation of the road by the driver and his crew. Put more precisely, by the notion of 'driving' I refer to the distinct ways, tasks and functions that the driver and his crew exert in order to make the journey. I will distinguish and describe three related driving practices in detail: path-making, course-making and the mitigation of hazards. Even though I describe these three bundles of practices separately, it should be understood that this is for analytical purposes only. In practice, they are inextricably linked with each other and complementary, and the distinction often depends on the perspective. The seating of passengers, for instance, may be described under the umbrella of passengering, as seen from the perspective of the passengers, but also as loading, as seen from the perspective of the driver. Only when conceived of as a joint overall practice of making the road can a clear understanding of travelling on the road be attained.

Loading

The first bundle of practices analysed here describes the process of packing, arranging and balancing the load. In this process, several diverging aims have to be considered and aligned with each other. The overall aim is to carry as much load as possible while keeping the lorry functional on the road. Yet the questions of how much load can be carried and according to which order the different items should be placed are usually subject to diverging opinions. In this sense, loading primarily means to negotiate different views and aims and, ideally, to align them with each other.

In the Sudan, as in other parts of the world, the legal frameworks related to axle-load and road safety define weight and size of the load.[3] The implementation of the regulations and the adherence to the law is overseen by the Sudanese highway and urban traffic police. These are generally present, or at least visible, on Sudan's asphalted roads; that is on highways and on urban roads. On these roads, lorries are compelled by the traffic police not to carry more than the standard maximum weight. The size of the load is also fixed and regularly observed by the police. Desert roads, by contrast, are not supervised by traffic police and thus allow for a looser observation of the law.

The desert roads are nevertheless not free from other agents of the state. Army and customs officials as well as a whole range of local and regional fee-collectors are usually stationed at entry points between states and localities, with each post imposing a certain tariff on the lorries that actually pass through (and do not circumvent) their checkpoints. Yet, compared to asphalted and

3 The Traffic Law of the Sudan (Government of the Sudan 2010) and the National Roads Protection Law (Government of the Sudan 1994) represent the legal framework that governs driving on Sudan's roads.

physically delineated roads, the fluidity of the desert road allows drivers to select tracks in order to circumvent and thus avoid state intervention into their business to a large extent.

The size and weight of the load on the hinterland road may vary widely from the legally prescribed norm. Flour, sugar, petrol, used furniture and even cars are piled up high on the lorry; and on top of the load, passengers sit with their luggage. The final size of the freight is negotiated by a multitude of actors and aims respectively; including passengers determined to get their (sometimes considerable) luggage to their destination, the forwarding agents bent on getting their full shipment dispatched at once and, not least, the driver cautious about reducing potential dangers stemming from the size and nature of the freight.

Loading According to the Manifesto

The management of the loading and unloading of contracted lorries is the responsibility of the forwarding agencies. Forwarding agencies are the bodies that regulate economic exchange on the road by channelling merchandise. Situated in the main markets of the capital, predominantly in Omdurman, these agencies function as mediators between merchants requesting commodities from Omdurman and the market in El Fasher. In addition, the agencies also facilitate the marketing of goods produced in Darfur. Each agency employs a manager and a group of loading workers who are responsible for the process. Regarding the financial aspects, the agencies operate according to set rules: for each item of freight, the forwarding agencies charge ten percent of its value as declared on the '*manifesto*', an official document specifying the items carried by the lorry. The 'manifesto' lists the cargo of the lorry for official use of customs, police and taxation purposes. This document is usually supplemented with a passenger manifesto that lists the names of the passengers, which is not an official document but is kept as a record at the forwarding agency.

The loading process takes into consideration the capacity of each lorry – its size, model, age, maintenance and modifications. Since the majority of the lorries have been modified for carrying more weight than they were originally designed to carry, the exact capacity of a lorry is available mainly through the knowledge of '*ahl al-lawârî*' (Ar. lorry community). The term refers to the people who are engaged in the lorry business, including drivers and their crew, employees of workshops, forwarding agents and lorry owners. The driver, based on years of experience and familiarity with his vehicle, is usually the one who determines the capacity of the lorry. Accordingly, his input to the loading process is the most decisive one and it is further complemented with the experience of the loading crew. The loading crew comprises two to four workers of the forwarding agencies who do the actual work of manually loading the lorry under the supervision of the driver. For experienced workers, the signature of the workshop engraved on the body of the lorry reveals useful clues about how much the lorry can take.

The driver and his assistant are always present during the loading process. The loading platforms in front of the forwarding agents' offices are equipped with chairs and Sudanese bed frames, on which the driver sits (or lies) while observing the loading process and ensuring that it is carried out orderly. The driver's main assistant is positioned on top of the cabin, keeping his eyes on the loading workers and listening carefully to the instructions of his driver.

The presence of the driver during the loading procedure is vital for the successful journey. By observing and directing it, the driver can better plan the journey ahead and, already here, prevent many potential hazards of the road. By attending to the loading, he can imagine and anticipate the performance and limits of the loaded vehicle on the road. Ultimately, such knowledge enables the driver to choose the most suitable routes, gauge the extent of the manoeuvrability of his vehicle and roughly estimate the duration of the journey. Mahajub, a rather novice driver, explained to me:

> Go and ask Ibrahim Al-Kumsanji [the forwarding agent], he knows that filling up the lorry is the task of his workers, but look at them, they just want to finish fast and start loading the next vehicle. They are *yaumîya* workers, they are paid per day and the faster they finish, the sooner they leave back to their homes. But it's me who is stuck to this lorry, if something goes wrong on the way with the vehicle or the goods, it is my problem. They will be sleeping on their beds and it will be me stuck in the desert.

The nature of desert routes – with their sand dunes and generally unstable surfaces – influences the way of placing freight on the lorry. Transporting freight on hinterland roads is characterised by a great diversity of items. Smaller items packed in boxes, sacks and barrels have to be balanced with the rather bulky freight, such as pieces of furniture, building materials and even small cars, all loaded together in an order agreed upon between the driver and the forwarding agents. Each of these items is placed according to size, weight, packing, value and fragility. Such ordering is intended to fit in as many items as possible on the vehicle. The logic of this order is decisively influenced by the aim of enabling, or maintaining, a certain latitude of manoeuvrability of the vehicle. For instance, when balancing the distribution of the load on a lorry, the loading workers try to place the heaviest load by the axles of the vehicle to make it stable on soft sand-tracks.

Loading practices are furthermore influenced by the value of each item. Certain items like sugar, fuel and flour are usually positioned at the bottom of the load, even if their weight does not usually qualify them as the heaviest pieces. This practice cannot be reduced to the intent of smuggling. These items should be invisible. Even if they were declared and taxed on the *manifesto*, these goods should be kept at the bottom of the load. Conceptualising this practice by legal terms does not make it possible for us to understand its real meaning and implications. For, in addition to perceiving these items as precious and strategic, they are also considered a source of trouble and danger. The road hosts several threats towards carriers of such items. The value of flour, fuel and sugar is calculated within the context of its scarcity on the roadside. Thieves, armed robbers, rebel groups or simply corrupt checkpoint soldiers are on a constant hunt for consumables and scarce items. As these guards usually receive their pay and provisions late (given that they receive them at all), they are often left to their own devices to obtain the staples. The first question asked at a checkpoint thus usually targets the staples: 'Do you carry any sugar, fuel or flour?'

An experienced driver answers according to his *manifesto*. Even if the staples are acknowledged in the *manifesto*, they are placed at the bottom of the freight. Hence, when asked for providing staples for the guards manning a checkpoint, the driver is able to negotiate the delivered amount and, due to the difficulty of unloading, he may in fact be able to keep the whole amount of the freight on board. If the presence of such strategic items was openly acknowledged, it would most probably incur risks from different sites, ranging from a checkpoint officer to a common thief at the truck stops.

Loading the takwîsha

During the loading process, the driver and his assistants engage in continuous negotiations about the shipment with the agency manager. While the driver tries to ensure a reduced amount of shipment in terms of size and weight, the manager pushes to fill the lorry as much as possible. The manager is looking to increase his commission whereas the driver aims to minimise the official shipment to ensure a space for extra load on top of the official load. This may include more passengers together with their luggage or other loads they may wish to transport and other unofficial cargo. While the official load is taxed and registered on the shipment's manifesto, the extra-load is not declared and thus tantamount to an added profit for the driver.

The overweight, hence the clandestine benefit of the desert road, has left its traces in the language of the people involved in the lorry business. The term '*takwîsha*' is referred to when speaking about the bulk of extra-load towering above the top edge of the lorry's body. Similar to the official load, the size of the *takwîsha* is used to assess the weight and manoeuvrability of a lorry. The *takwîsha* is also considered a manifestation of the expertise, skills and experience of the driver to navigate and handle his car. A large *takwîsha* is perceived as a source of pride. And the larger it is, the higher the status and the potential financial benefits that the driver can claim.

In the Sudan, as well as in other African countries, travelling long distances figures prominently in the cycles of commodity exchange. Travellers often bring along piles of agricultural and husbandry products as luggage. The fluid nature of the road offers petty traders in particular welcome opportunities while on the other hand, compelling them to accompany their goods in person instead of simply sending them through a forwarding agency. The unregulated roads offer opportunities for reaching the markets of the capital with a limited level of state control and taxation. While travellers coming from the capital and central areas are usually loaded with industrial commodities such as electronics, spare parts and fashion items, rural traders bring with them locally produced goods. Both groups undertake the journey with a view not only to obtain more favourable prices in comparison to the local prices but also when compared to other ways of transportation that are regulated and taxed. The practices of loading and the fluidity of desert transport in general, combined with the opportunity of travelling alongside and thus supervising their transported goods, render the rather tedious experience an attractive option. From this perspective, it is in fact 'safer' to take the uncertain desert road than risking complications like checkpoints, heavy traffic and axle-weight limits on the highway.

Passengering

Lorry travellers are nonetheless not exclusively small merchants transporting their goods. There is in fact a broad diversity of travellers, including common 'commuters' (who, unlike the merchants, have less luggage to transport) as well as students on vacation and people on their way to try their luck in the gold mines. Once a traveller arrives at the lorry and pays the travel-fare, the quotidian order of being on the ground is replaced by new temporalities, setting time and space into new coordinates that are situated in the practices and particular socialities of travel. The everyday time passing by between meals, work and prayer times is replaced by a several days long, continuous journey that is only interrupted by unannounced truck stops and sudden breakdowns.

As stated above, passengers have an important role in the making of the desert journey. Their practices, such as waiting and seating, are covered by the term passengering. Despite the centrality of passengers to the overall travelling processes, their presence remains largely unacknowledged in the respective literature. For instance, Jain (2009) points out that understanding of passengering within official discourses tends to minimise the role of travellers in the making of the journey. Passengering is also being analysed as an artful social agency (Vannini 2011) or a set of performances (Adler 1989).

Waiting

At the beginning of the journey, the view of the lorry being loaded provides for the passengers a kind of visual time similar to that of a sandglass. As piles of boxes and sacks of merchandise are being loaded at the premises of the forwarding agency, each level of boxes stipulates a reduction in time of waiting for the departure. Passengers eagerly observe the loading while their minds and conversations aim at an estimation of the time left. The process of hoisting the varied goods onto the back of the lorry can surprise even seasoned travellers, particularly due to the excessive volume of the load. While experienced passengers might indeed be able to acquire a mental picture of the overloading of these desert-going lorries, the height of the load on which they are going to sit will, more often than not, still surprise them. Most lorries are loaded up to the top of their cabin, then the *takwîsha* is piled up, and on top of that passengers have to find a space to occupy while jerry cans, pieces of furniture and other bulky items are tied to the back and the sides of the truck.

The waiting time might seem wasted and empty (Watts and Urry 2008; Gasparini 1995). Experienced travellers, however, keep watch of the process well aware that the time spent watching represents the critical hours for their upcoming journey. As I witnessed during my first trip, the loading process is crucial and the seasoned travellers can assess the length of the journey by it. As a middle-aged man commented: 'This lorry is too heavy, they are overloading it and we will get stuck on every sand dune on the way. I better call my family that I will be a day, maybe even two days late.'

Loading allows for initial interactions between the travellers, which lay the foundations for their later sociality. It allows for imagining the ways of the travel and for sharing and discussing the options. Being present during the loading process becomes literally the only way to travel, as passengers who do not wait are likely to miss the departure. Even if the departure is delayed for the next day, sleeping close to the vehicle is the only guarantee to be aboard the next day.

This seemingly empty time is even more crucial for the next phase of travelling. Attending and watching the loading of the lorry allows for locating possible spaces to occupy on the top of the load. When passengers board the vehicle they strive to occupy the 'least uncomfortable spot'. For example, sitting or lying on top of a metallic bar with sharp edges would mean a painful experience for the next 800 kilometres. Wrong positioning in the beginning can make even shorter trips painfully long. Most of the passengers, however, travel the whole journey and shifting place is rather frowned upon as this means a disturbance of the social order on the overcrowded lorry.

Seating Practices

The quest to seize a less (or 'the least') uncomfortable space on the back of the lorry entails more than attending the loading or trying to avoid back pain. Each position creates a certain travel experience. Sitting in the front or at the rear of the lorry, or on the side of it, is not only decisive in terms of the scenery of the monotonous desert landscape, it is also decisive regarding the degree of exposure to the environment. Sitting at the edges of the lorry increases the exposure to wind, while sitting by the rear of the lorry results in constant suffering from the clouds of dust stirred up by the wheels. During the trip, the exposure constantly shifts according to different speeds of the lorry, time of the day and condition of the tracks. Consequently, the preferences of the passengers to occupy certain positions also change. During the daytime, passengers prefer to observe the environment, whereas during the night they rather need cover from the cold wind and more space to stretch their legs under their blankets.

Alternative levels of exposure differentiate the experiences for the travellers. When on the move, potential conversations between the passengers are muted by the constant wind. Passengers seated in a central position are better able to communicate, while passengers at the edges of the lorry are just busy protecting their faces from the dust. Being 'at the edge' has a literal meaning on the lorry. Blinded and deafened by the blowing wind, one is constantly in search of something to grasp to protect oneself from falling.

The spatial arrangements on top of the lorry are, however, never stable, but contested and negotiated. The limitations of space and the desire for a more comfortable position continue to influence the order on the lorry. Competition over places entails pushing and pushing back, complaining, faking pain and even quarrelling. On the other hand, such conflicts are habitually mitigated through sharing conversations and snacks. In order to have a smooth journey, competent passengers have learnt to balance their needs and comforts.

Consequently, a delicate balance is reached between the search of a less uncomfortable position and maintaining the social codes valid on the ground. This is more visible when women and men have to share the top of the lorry. In these situations, women are automatically offered a central position on top of the load, while men take the exposed positions at the edges of the vehicle. Cabin travellers represent a different class of passengers. The cabin of the lorry is usually equipped with two seats in addition to the driver's seat, although sometimes three are squeezed in. Some of the more recent lorry models even have closed cabins. The closed windows in the cabins represent a luxury – travellers are isolated from the outside weather and provided with a panoramic view of the road and the surroundings. The convenience of being a cabin passenger is highly appreciated. Accordingly, it also costs more. While waiting for the loading of the lorry, travellers wait in the same area close to their lorries. During this process, there is no indication of who is going to be a cabin passenger and who will travel through the desert on the top of the lorry.

The particular social space created by passengering on the back of the lorry is also manifested in the talks and in the topics discussed during the journey. Travellers tend to avoid speaking about who they are, where they are from and where they are heading to – leaving behind the established norms and patterns of normal life 'on the ground'. Popular among the travellers on the back of a lorry are stories about the road and about previous experiences. The landscape provides additional opportunities for non-controversial topics, and the most common questions relate to the current position and the road ahead, such as 'where are we now?' and 'where is the next truck stop and when will we arrive there?' Such conversations are critical in consolidating the travellers into a travelling community and in acquiring competence in the practices of passengering.

As the course of the journey itself is determined by passengers' destinations, they contribute directly to the making the journey and thus to the making of the road. Even a passenger with little luggage is taken into consideration when setting the course of the journey. Before the beginning of the journey, the driver tries to maintain a short course for his travel with the least amount of detours towards his destination, although, as soon as a new passenger is accepted to board the lorry, the course will be modified to include his destination.

Through repeated travels, passengers acquire certain competences in assessing the skills of the driver. Experienced travellers are able to identify certain landmarks, ways of driving as well as sharing their ideas when clinging to the top of the lorry. Assessing the driver's skills when confronted with a sand dune (Ar. *qauz*) is an integral part of passengering. For instance, whether a driver decides to push forward immediately or whether he deflates his tires before entering the dune reveals much about the extent of the driver's skills to the seasoned travellers. Passengers try to deduce from the skills of the driver the estimated time of arrival and engage in lively discussions about his overall performance. In doing so, storytelling and the sharing of previous experience provide for opportunities to acquire competences in the practices of passengering.

Revealing interplays between practices of passengering and driving occur both on and off board. The need of being an agreeable travel companion – by avoiding conflicts, making alliances and bonding with others – is supplemented by the complementary need to safeguard one's seat and one's status. During the first stop, the physical partition, and thus social differentiation, of the travel community is made apparent when the passengers disembark on the sand. At the rest stop, cabin passengers are invited by the driver to his favourite '*qahwa*' (Ar. coffee-house), whereas passengers on the back of the lorry have to look after themselves.[4] The first stop usually occurs just a couple of hours after departing, the atmosphere is very smooth and calm, people are very friendly and talkative, the guests of the coffee houses are happy and laughing and the overall feeling of relaxation is present. In comparison the stops occurring later, especially in the area between Jabrat ash-Shaykh and Soderi, have a distinctively worse atmosphere. After hours on board the passengers and lorry crew are exhausted, annoyed and wretched. The bad temper is omnipresent, the typically friendly coffee house staff are also aggravated when rebuked by travellers: 'Is it them whose bones were shaking for hours? My money is good for them but this coffee is not worth a dime'. To such talks, the ladies running the coffee houses often silently swear: 'Why on earth did God throw me into this place?' The amount of time devoted to resting is the driver's decision. Yet passengers try to interfere, not least because while they have no chance to make their voice heard when on the back of the lorry, the rest stop provides them with the opportunity to address the driver with their claims. Despite the fact that it is the driver who is actually working, it is typically he who is pushing for a shorter break, while the passengers request more time to catch their breaths. The driver prefers to arrive as early as possible in order to conduct as many journeys as possible. On the other hand, the distress of travelling discourages even the most urgent travel obligations of the passengers who try to seize any opportunity for not being on the lorry.

4 For further illustration of the sociality in Sudanese truck-stops and coffeehouses, see Beck 2013

Driving

As already delineated above, despite being referred to as a road, the Forty Days Road is in fact not one road but rather a bundle of desert roads that serve different villages and towns while running roughly from east to west (and back). Each of these roads usually consists of several vehicle tracks that are only suitable for a certain type of wheel. The tracks split and merge again according to the terrain with a width that varies between 10 and up to500 or more meters.

To the trained eye of the driver, the road constitutes more than just an undifferentiated landscape. He chooses his course through a whole range of different landscapes, towns and villages, topographies and soils. For the driver, the succession of different landscapes constitutes the progression of the road, in that its landmarks roughly define his ways, yet the decisions over which track to follow and which to avoid rests on skills that are not easily described. These decisions have to be made in short moments that leave little room for thorough evaluation, and they are based on vague assumptions as to the state of the tracks ahead. Driving therefore consists of, firstly, a kind of path-making that can be conceptualised as comprising of largely ad hoc and tacit decisions responsive to the uncertainties of the road and, secondly, of a kind of course-making that relates to the succession of landscapes and the tracks taken to reach a specific destination on the way to the west. In the following, I explore the practices of driving through these spatial and temporal practices applied by the driver in his engagement with the road.

The cardinal directions, north, south, east and west, are typically fundamental sources of orientation when on the move. However, as a general rule, when travelling on the Forty Days Road, the drivers do not rely on these directions. Bearing in mind the general direction of either west or east (depending on their destination), knowing the geography and relying on the movement of the sun, drivers are able to maintain the general sense of direction. In some areas, these directions change slightly and the lorry is heading north or south, yet such changes are barely recognised by the drivers. The references to directions within the context of the road are not geographical – references to east or west are replaced by terms of forward '*qiddâm*' and backwards '*warra*'. Both terms correspond to *qibla* reckoning where *qiddâm* points to Mecca, the direction Muslims face while praying.

For the drivers, the sense of a moving vehicle is reversed. Seen through the windshield, the road and its environment are rather moving towards the vehicle instead of the vehicle moving forward. Instead of approaching specific objects, the drivers, while way-finding, see the objects as moving towards them. From this view, the driver creates an image of the movements and the positions within the surroundings. Driving in this context can be defined as the deployment of various skilful methods of acquiring knowledge about directions and whereabouts while on the road – where the spatial configurations displayed through the windshield of the lorry provide a way of understanding the environment (see Golledge 1992).

Driving depends on the skilful tracing of a set of landmarks through the windshield. On the road, the success in tracing the landmarks is the primary method of producing a sense of location and orientation. The spatial elements of these landmarks as objects of nature are not as important as the spatial meanings the driver associates with a certain location. A small bush, a fallen tree, an abandoned tyre or a rocky hill can all constitute a possible landmark for passing drivers. Each driver collects and memorises an individual set of landmarks from his own passing perspective to assist him in identifying his usual routes. Although experienced drivers usually learn the local names of landscape features from knowledgeable passengers and from conversations at the rest stops, these are not their only meaningful landmarks. They also establish landmarks from their own experience and from their own perspective through the windshield and thus create their individual set of landmarks that is only meaningful to them. Some drivers even assert the need to associate the shape or the order of landmarks with figurative memories. For instance, one driver mentioned to me that an abandoned tyre we passed by was from his own vehicle, and that during the last three years he passed through that particular location, the tyre was still in its place. Another driver recognised a small bush on the road, referring to it as 'the snake's bush'. When I asked him what made this particular bush a landmark and whether it was also recognised by other drivers, he answered that it was only known to him and the driver whom he used to assist. They gave it this name because of a snake they saw there when their lorry broke down near to the bush once.

Path-Making

The physical challenges of driving on a desert road materialise in particular on the long stretches of the soft and unstable sand-tracks. As Thomas (1952) describes in a contribution to the geography of Saharan travel: 'the road ends where soft areas, no different in general appearance from firmer surfaces, are encountered. The tracks fan out making a trail hundreds of yards, or even many miles in width.' He goes on to describe driving on these tracks: 'the first vehicle over such a surface breaks the crust, so that a truck following the same track is likely to sink into the sand and become stuck. Therefore, each driver attempts to find a new course, but close enough to the other to avoid getting lost' (Thomas 1952: 273).

In order for drivers to avoid getting stuck in the sand, or losing the balance of their vehicles, they select and re-select suitable tracks. I refer to the practice of selecting the right track for the vehicle as path-making. Depending on the particular type and weight, each lorry can only navigate a particular set of tracks on the road. For instance, lorries with a dual rear axle require different tracks than single axle lorries; fully loaded lorries have a different lane to follow than less loaded ones. For the knowledgeable eyes of the drivers the intersecting lorry paths, depicted below, communicate spatial information just as effectively as the map at the beginning of the chapter.

While driving, each track is a source of rich knowledge for the drivers, with its specific imprints bearing the marks of vehicles that have successfully passed through it. By displaying the history of the traffic and the type of vehicles that have travelled on its surface, the track with its imprints provides information physically written on the road that can be deciphered by a competent driver (Figure 8.2). The road itself, with its routes and tracks, is an outcome of long standing practices of travelling on its paths; or as Ingold puts it: 'Remembering is not so much a matter of calling up an internal image, stored in the mind, as of engaging perceptually with an environment that is itself pregnant with the past' (2000: 189).

FIGURE 8.2 *Intersecting paths on the road*
SOURCE: R. WADELNOUR, 2013

Path-making involves more than visual scrutiny of the road. Other sensory elements are also crucial to the making of paths. While seated behind the steering wheel, the competent driver is able to assess the reactions of his vehicle to the terrain of the road by feeling the movement of the vehicle through his body.

It has been claimed above that the practices of making the desert journey are related to each other. Desert travel is basically a collective endeavour and the different actors contribute to its success in different yet related ways. Just as passengering and loading, as well as loading and driving represent interconnected bundles of practices, so are practices of passengering and driving interconnected. Night driving illustrates the interplays between the driver and the passengers. Night driving makes clear how the solitary, individual practices of navigating are complemented by a set of social practices established through the interaction between the travel community that emerges from the journey. Night driving, with the darkness and minimal visibility, presents a veritable challenge to the orientation skills of the driver, as it hampers the track-reading and way-finding practices based on visual recognition. Although a plight, driving at night is nevertheless the preferred mode of desert travelling. The lower temperatures during the night, when temperatures rarely go above twenty degrees Celsius compared to the common temperatures of over forty degrees Celsius during the day in the hot season, are comfortable both for the drivers and their vehicles and are believed to facilitate driving in the sand by consolidating the sand dunes, thus reducing the risk of getting stuck in the sand.

After driving for most of the day, drivers tend to become less focused during the night. Accordingly, their need to rest and sleep intensifies. To fight the pervading fatigue, drivers try to stay alert by speaking to the cabin travellers. Talking in loud voices and about controversial topics becomes compelling during the night. As the cabin travellers have to balance their own need for sleeping with the risk of the driver falling asleep, they are urged to fuel conversations as a means to make more miles in the cooler desert night. In contrast to the conversations during the day, which tend to centre on 'easy' topics discussable by anyone anywhere, night conversations commonly trespass social conventions and boundaries. While the conversations during daytime are centred on weather, prices in the market, guessing the distances and next stops, the accentuated anonymity of the night creates a new setting. In one of those talks, a driver referred to as Ash-Shaiqî (referring to a tribe from Northern Sudan) told me how sometimes one's name can work against a person on the road:

> You know they call me Ash-Shaiqî. But I am not even a Shaiqî! I swear to God, I am from the Kabâbîsh [a tribe from Kordofan]. I lived in Merowe,

> I did marry a Shaiqî woman but I am not a Shaiqî. I lived most of my life with them, I speak like them and yes I sound like them, but I am not a Shaiqî; people just call me like this. Once I was driving along this road and we got stopped by a group of robbers. They were quite decent and polite, just robbers. While they were busy taking valuables from the passengers, my stupid assistant referred to me as Ash-Shaiqî. At that moment a rifle was pointed at me directly and one of the robbers asked: 'So you are a Shaiqî! You are not leaving this place! This is it for you!' I started explaining that I am actually not a Shaiqî and that I am a Kabbâshî like them, but he did not want to believe me at all, so I spoke for my life with that gun pointing at me. As you can see I convinced him, I am really a Kabbâshî. I was lucky, but hey that was close, such a stupid person! Me a Shaiqî? The things people say.

Course-Making

As a crucial aspect of driving, course-making is a means of identifying the temporalities of the journey. Breaking down the unconfined path, course-making is the practice of breaking each journey into shorter, bearable trips. From the perspective of the driver, a full trip includes the journey from Omdurman to El Fasher and back to Omdurman. The '*mushwâr*' (Ar. journey) is the measuring unit for the drivers. On average a *mushwâr* lasts nine or more days. Drivers try to conduct two to three trips each month. Within each trip, the journey from Omdurman to El Fasher is considered the main task that is broken down to smaller units. The longer the journey, the more difficult it becomes. The round trip between Omdurman and El Fasher is considered as both the longest and the most challenging trip, and the longer and more challenging trips generate more profit for the driver.

Once on board, the particular temporalities of the journey take on determining roles. Measuring units of time and space merge and become confused. Distance is not measured solely in kilometres and hours or days, but also in hunches, guesses and feelings. The bush on the side, the desire to arrive soon, the annoying voices on board and the prospect of a rest stop all become part of course-making.

Desert routes entail certain rhythmic temporalities of making the journey. Most drivers tend to avoid travelling during the afternoon. Besides the blinding sun and the scorching heat that endangers the lives of travellers, the high temperatures can also cause serious problems for the vehicles. As a result, most of the course is conducted during the morning, evening and at night. During the stops in-between these passages, the travel community is able to rest, eat or sleep at truck-stops scattered along the road (see Beck 2013). These breaks are a joint decision of the travelling community. Certainly the driver is holding the wheel and arranging the path, yet the decision to stop is largely influenced by the insistence of crew members and passengers, as well as by the appearance of a destination on the way or by the heat becoming unbearable. Hence, during the journey, multiple stops are made, some are missed and many are delayed according to the driver's determination to ignore the demands of his passengers.

Course-making does not involve announcements; rather it is the art of the untold. Most drivers abstain from declaring their course of journey or even their planned next stop. Avoiding sharing such information is not simply a security strategy; the purpose of withholding information needs to be interpreted in its multiple layers of meaning. The uncertainty of the road, with its ever-shifting nature, challenges the word of the driver. And if the driver is seen as not keeping his word, he risks losing his authority and may eventually need to surrender even more of his decision-making power to the other voices on board. Rather than laying their plans open for scrutiny, most drivers therefore choose to keep their paths and routes to themselves.

Mitigating Hazard

Security plays a crucial role in shaping the road regime. The drivers on the Forty Days Road, while facing the challenges of traversing vast territories with ever-changing features, have to deal with further challenges, such as unexpected mechanical defects, the dangers of losing track of the route and getting lost or ambushed. As the driver's main task is to steer his lorry and the load safely and timely through the desert, hazard mitigation is part and parcel of the practices of driving. It has a clear interrelation with course-making, as to incorporate security concerns in course-making is a fundamental competence for the driver. The extent to which these uncertainties figure in the drivers' overall concerns becomes manifest in their extensive use of protective amulets and spells, which is a common practice in the Sudan. Leather-made amulets that are believed to protect from certain harms decorate their bodies and their vehicles. Certain amulets are believed to shield against bullets, others against knife attacks, while others are used to protect the vehicle from serious mechanical damages. The levels of uncertainty related to travel on the Forty Days Road can also be traced to the ways in which the communities on and along the road commonly refer to the road. For example, since the escalation of hostilities in the region in 2003, the term *darb al-hawa*, in Sudanese colloquial Arabic 'the road of trouble', has come to figure prominently as its reference. Ever since, stories about ambushes on lorries, kidnappings and armed robbery on the road are proliferating. In order to counter the growing insecurity, the Sudanese armed forces and its ally tribal militias have established the practice of armed convoys for commercial vehicles crossing into Darfur.

The closer travellers come to the borders of Darfur, more imminent feelings of danger rise. Yet while mitigating these dangers, travelling in a convoy also creates new challenges. Travelling in a convoy means that the established order of the travelling community on the lorry is fundamentally disrupted. The discretion to determine his own path according to his own assessment of the road, which formerly was at the heart of the drivers' practices of course- and path-making, is considerably reduced as the driver becomes obliged to follow a path only a few meters wide. Restrained by a train-like order in which a vehicle closely follows another vehicle, connected together for protection and security, the driver loses his capacity to closely attend and react to the terrain. Collisions between lorries, which have never been a problem in the vastness of the desert, become an eventuality in the forced proximity of the convoy.

When boarding the already crowded back of the lorry, the convoy's militiamen claim the best places, thus disrupting the spatial arrangements created in the beginning of the journey and which have been consolidated into the particular order of the journey during the trip. Overcrowding, uncertainty and pervasive insecurity foster feelings of discomfort between the passengers and the newcomers. The equipment of the convoy's militia members – usually AK-47 and G-3 rifles – further add to the tension of the situation, deepening the sense of insecurity.

As argued by Beck (2013), truck stops aim to confine the roadside community from the road while simultaneously providing a connection between road and roadside. They represent legitimate spaces for contact between the road and the roadside, while other contacts are discouraged and sometimes even feared. Similarly, escorted convoys aim to provide a safe passage through the territories of roadside communities. In this sense they are also mediating institutions which channel the conduct between road and roadside. Through the convoy's guarded and scheduled itinerary, the political leadership of roadside communities is able to maintain a higher level of control over movement within their territories.

Conclusion

Having identified some of the experiential aspects of travelling the Forty Days Road, this chapter has distinguished three bundles of travel practices that together contribute to the making of the road; namely the practices of loading, passengering and driving. This broader differentiation – as well as the more detailed differentiation into practices such as formal and informal ways of loading, waiting, seating, way-finding, course-making, path-making and mitigating hazards – is for analytical purposes. In making the journey, all three bundles of practices are inextricably linked with each other and complement each other. The analytical differentiation, however, allows for highlighting practices that conventionally tend to be overlooked. This pertains in particular to passengering, and thus to the *active* engagement of passengers in the making of the journey. For, conventionally, passengers are ascribed a passive role only. They are transported. But the lesson from the Forty Days Road is that even though they are transported on the back of the lorry like other cargo, they still actively engage in the organisation of the journey. This points to the negotiated orders of travel in Africa, which is also highlighted in other contributions to this volume (see in particular Musch and Stasik).

By conceptualising the Forty Days Road as being produced through travel practices, this chapter furthermore opens an alternative perspective on the making of roads more generally. Roads that are not travelled tend to vanish. Only roads that are actually used are established and consolidated as roads. Hence, while roads do provide a foundation for traffic – and this is the conventional perspective – in the case of the Forty Days Road it is the traffic itself that brings them into being. When seen from outside the confined perspective of road planning, one may well ask whether this is not the actual sequence of events by which roads are brought into existence in Africa: first the traffic, then (eventually, if at all) the asphalt. In this regard, the case of the Forty Days Road could thus speak to much larger issues than the Sudanese hinterlands.

References

Adler, Judith. 1989. 'Travel as Performed Art.' *The American Journal of Sociology* 94 (6): 1366–1391.

Ashmore, Paul. 2013. 'Slowing Down Mobilities: Passengering on an Inter-War Ocean Liner.' *Mobilities* 8 (4): 595–611.

Beck, Kurt. 2009. 'The Art of Truck Modding on the Nile (Sudan): An Attempt to Trace Creativity.' In *The Speed of Change: Motor Vehicles and People in Africa 1890–2000*, edited by Jan-Bart Gewald, Sabine Luning and Klaas van Walraven, 151–173. Leiden: Brill.

Beck, Kurt. 2013. 'Roadside Comforts: Truck Stops on the Forty Days Road in Western Sudan.' *Africa* 83 (3): 426–445.

Gasparini, Giovanni. 1995. 'On Waiting.' *Time and Society* 4: 29–45.

Gerstenberger, Heide and Ulrich Welke. 2004. *Arbeit auf See: Zur Ökonomie und Ethnologie der Globalisierung*. Münster: Westfälisches Dampfboot.

Golledge, Reginald G. 1992. 'Place Recognition and Wayfinding: Making Sense of Space.' *Geoforum* 23: 199–214.

Government of the Sudan. 1994. *Qânûn himâyat at-turuq al-qaumîya li's-sannat 1994* (Ar. National Roads Protection Law 1994). Accessed July 18, 2016. http://www.moj.gov.sd/content/lawsv4/7/18.htm.

Government of the Sudan. 2010. *Qânûn al-murûr li's-sannat 2010* (Ar. Traffic Law 2010). http://www.parliament.gov.sd/ar/activity/view_law.php?law_id=311.

Hänsch, Valerie. 2009 *Sifinja: The Iron Bride*. Documentary DVD, Germany, 70 min.

Hutchins, Edwin. 1995. *Cognition in the Wild*. London: MIT Press.

Ingold, Tim. 2000. *The Perception of the Environment: Essays in Livelihood, Dwelling and Skill*. London: Routledge.

Jain, Juliet. 2009. 'The Making of Mundane Bus Journeys.' In *The Culture of Alternative Mobilities: Routes Less Travelled*, edited by Phillip Vannini, 91–110. Farnham: Ashgate.

Klaeger, Gabriel. 2009. 'Religion on the Road: The Spiritual Experience of Road Travel in Ghana.' In *The Speed of Change: Motor Vehicles and People in Africa, 1890–2000*, edited by Jan-Bart Gewald, Sabine Luning and Klaas van Walraven, 212–231. Leiden: Brill.

Kleinert, Martina. 2009. 'Solitude at Sea or Social Sailing? The Constitution and Perception of the Cruising Community.' In *The Cultures of Alternative Mobilities: Routes Less Travelled*, edited by Phillip Vannini, 159–175. Farnham: Ashgate.

Thomas, Benjamin E. 1952. 'Modern Trans-Saharan Routes.' *Geographical Review* 42 (2): 267–282.

Vannini, Phillip. 2011. 'The Techne of Making a Ferry: A Non-Representational Approach to Passengers' Gathering Taskscapes.' *Journal of Transport Geography* 19: 1031–1036.

Watts, Laura and John Urry. 2008. 'Moving Methods, Travelling Times.' *Environment and Planning D: Society and Space* 26 (5): 860–874.

CHAPTER 9

Teda Drivers on the Road between Agadez and Assheggur

Taking over an Ancient Tuareg Caravan Route

Tilman Musch[1]

Nous sommes libres, nous sommes les maîtres du Sahara. With these words, young Tubu Teda defined themselves when talking to me. They are the drivers of Toyota Hilux pick-ups who transport West African migrants and who 'smuggle'[2] goods on the road from Agadez (Niger) to Sabha (Libya) and back. With these words, they also expressed their claim of 'ownership' over a vast desert space.

The Teda live in northeastern Niger (Manga, Kawar and Djado), northwestern Chad (Tibesti) and southern Libya (Fezzan). They may have some common origin in the Tibesti, where their 'supreme chief', the *derde*, still lives today. Many Teda are camel pastoralists, but they also own and exploit date palms; they may do gardening or work as merchants, in particular in Libya.[3] Formerly,

1 I express my gratitude to all the travel and 'sedentary' communities who contributed to this research: the Teda drivers who shared with me the journeys through the Tenere; other Teda who always received me with the great hospitality which is so proper to peoples of the desert; inhabitants of Dirkou, Bilma, Agadez and other places of Niger who showed no less hospitality in receiving me; the soldiers, customs officers and other state authorities who supported my research in their own way; the migrants, who gave me some important insights into their dreams of a better life; and not least, the Tuareg with whom I have also been working for several years. I also thank Kurt Beck and his team who edited this volume for accepting my paper and for contributing to improve it.

2 The quotation marks should relativise the negative connotation of the term. In fact, a driver who considers himself as the master of the desert and who considers the Sahara as his homeland divided by arbitrary (colonial) borders is not 'smuggling' but only transporting goods. The current political mainstream is considering the transportation of migrants and 'smuggling' as 'illegal'. It is for the same reasons that I will refrain from mentioning the names of my interlocutors.

3 Despite some seemingly 'sedentary' activities Teda engage in, most of them always maintain the option of mobility. Thus, a garden can be abandoned or left to a family member for several years (until the owner 'returns'), and a merchant may travel to acquire merchandise. The most prominent example for 'agricultural' mobility are the date palms – many Teda families own plantations in several places at far distances from each other (for example Kawar and Djado), whereas the owners themselves may stay in a third place (for example Manga) with

 | DOI 10.1163/9789004339040_010

the Teda controlled the caravan trade between the Manga, the Kawar and the Fezzan (Fuchs 1983: 131–132), and they have conserved an excellent knowledge of the topography of these regions.

One may be surprised that the western part of the space that the Teda cross on their ventures (see illustration 1) includes tracks used and perhaps established by the Tuareg's famous salt caravans leading through the Tenere desert. The Teda did not participate in these caravans which were almost exclusively organised and avidly protected by Tuareg. Furthermore, most Teda with whom I discussed the matter of travelling through the Tenere conceded that, for them, crossing this space was not without risk as some people considered them as foreigners. Crossing the Tenere desert and coping with the communities of its (former) 'owners' can thus be a challenge for the Teda drivers. They seem to manage this challenge rather well: they take over the desert space by means of their extremely high mobility and their ability to appear and disappear. In this chapter I explore how the Teda take over the ancient caravan track, focusing in particular on the ways in which they combine long-standing experiences of travelling with modern means of driving. I also show how they come to shape 'their' road by interacting with other travel communities.

Theory, Methods and Travel Context

Most of the Sahara, with the exception of oases, is hostile to human life. Nevertheless, humans are present in the desert, and they claim, appropriate or use its space. As I will show, the traveller from Agadez to Assheggur does not cross a no man's land but rather spaces, which different groups consider as 'theirs'. What does it mean exactly to 'claim', 'appropriate' or to 'use' space in the context of the desert?

Due to the hostility of the environment, claiming, appropriating or using desert space can only be realised by movement. Caravans crossing the Sahara, for example, have to do it as rapidly as possible otherwise they would lack provisions, which are carefully restricted so as not to overcharge animals (Chapelle 1982: 114; Fuchs 1983: 174–178). In a contemporary context, Klute (2009b) shows that the control of time in desert wars is calculated, among others, according to the rapidity of a vehicle. This is more important than the control of space, emphasising at the same time that the two concepts are closely linked. In the context of deserts, crossing the space is essential, as staying in it would

their livestock. Family members (mostly women) will travel to the palm-yards only during harvest.

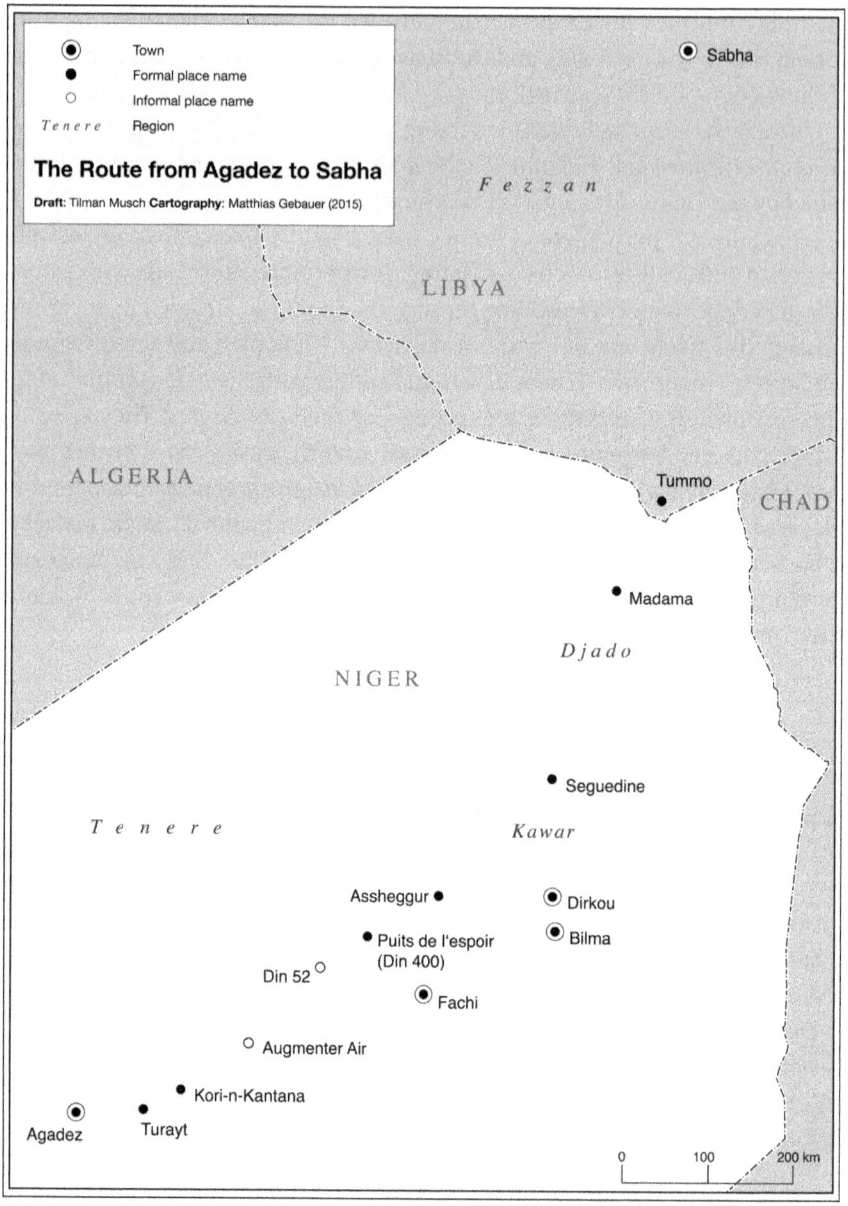

MAP 9.1 *The road between Sabha and Agadez*

be futile. One can thus assume that 'competing for space' refers to owning the better means to leave a place quickly and to arrive at another within a short space of time. In a very concrete sense, such means are motorised vehicles or camels. But roads or tracks are also means that make travel possible and when we assume that the important thing in the desert is to move and not to stay, then mastering space means the competition for and the shaping of roads.

The term 'road' in the present contribution refers in a wider sense to a travelled way, as, for example, in the case of the Forty Days Road (Wadelnour, this volume). Such a 'road' is of course different from many 'modern' roads which designate a clearly delimited, built and often bituminised way. The road as discussed in this contribution consists of tracks following mostly the same direction and which are loosely delimited by dunes, rocks or other non-navigable terrain.

Formerly, the camel, being a well-adapted means to cross the desert, was widely used. Nowadays, the Teda, having 'appropriated' a new means of transport, use Toyota Hilux pick-ups. According to Beck (2002), appropriation is a dialectical creativity-based process between the appropriated thing and the appropriating agent, which also implies changes in the life or values of the appropriating person or society (see also Beck 2009; Hahn 2004; Klute 2009b). Considering the societal changes in the process of appropriation, what could this mean for the Teda drivers?

I propose to concentrate in particular on their perceptions of space and on their making of the road. I will discuss how features of the Teda's mobility are now adapted to a context of modernity, and how a road is newly shaped according to the Teda's understanding of travel.

The making and perception of the road is not only influenced by the changes that Teda drivers undergo through their appropriation of a 'modern' means of transport – the road is also shaped by the relationships between different travel communities, namely Teda, soldiers, migrants and Tuareg. In his study of sociality among road and roadside communities, Beck (2013) shows how hospitality towards strangers can be altered by changes of space and time – the modernisation of traffic and truck-stops on the Forty Days Road in Sudan has led to a professionalisation of hospitality and to the segregation of travelling strangers. In the following, I will show how sociality among travel communities can influence the perception and the shaping of the road. At the same time, I will discuss how relationships between communities also depend on changing manifestations of crossed space.

My contribution is based on ethnographic fieldwork among Tubu Teda, which was carried out in Niger in the beginning of 2014. During that time, I had the opportunity to travel twice with Teda transporting West African migrants on Toyota Hilux pick-ups through the Tenere desert, from Agadez towards Dirkou and back. Entering the Fezzan however was not possible due to insecurity in southern Libya.

During my travels, I conducted research through participant observation and informal discussions (in French), in particular during breaks when the drivers had more time for conversations. In fact, travelling offers a productive moment for ethnographic research because the travel community is somehow 'disconnected' from every-day social relations (Spittler 1998: 30-31). Doing research on Saharan roads essentially means working in what Scheele (2012: 20) refers to as 'the "in-between" spaces'. These spaces maintain and link travelling and roadside communities that are 'stretched out' over vast space.

I also carried out conversations in the towns of Bilma and Dirkou and in the villages of the Kawar. Rather than interviewing 'informants', I exchanged informally with road companions who appeared to be pleased that a European researcher had shared their travel with them. This exchange on an 'equal' level made it possible to discuss freely matters of travel and other topics. I also copied data from the drivers' GPS and asked them to draw mental maps (see Lynch 1976) for me of their tracks (see Fig. 9.3).

I had not initially planned to travel with migrant-transporting Teda. I had intended to conduct research in the Kawar and was just searching for a means

FIGURE 9.1 *A Teda driver at Turayt on the way back to Agadez*
SOURCE: T. MUSCH, 2014

to reach there. In Agadez, my friends secured a place for me in a vehicle that was supposed to be in good condition and with a 'serious' driver. The vehicle, however, still broke down less than 100 km after leaving Agadez and just before reaching the military post of Turayt. It was late in the evening and the driver sent me to the post in order to search for help. At the post, the soldiers were very polite and received me well. They told me that it would be no use to go back to my car as its repair was uncertain, and they proposed two solutions: either to spend the night at the post (they kindly offered me a mattress and blanket), which was a rather comfortable solution but without a realistic perspective to continue the next morning, or alternatively I could continue in another vehicle on the convoy they were about to arrange. The soldiers and I debated and decided on the second option and they stopped a Teda whom they considered to be a 'good driver'. I continued my journey with him and women migrants from The Gambia[4] on the back of a Toyota Hilux, without my warm clothes for the night, sunglasses or my turban, as all of these were left behind in the broken-down vehicle. Only five days later did the gendarmes bring my luggage to Bilma. The Teda driver with whom I travelled until Assheggur refused any payment. For him, it was a question of honor.

At Assheggur my driver wanted to head straight northwards in order to avoid approaching the bigger towns of the Kawar. This explains why I continued to Dirkou in a vehicle of the national customs service. On my way back from Dirkou to Agadez several breakdowns occurred and I had to use three different cars altogether. The fact that both journeys had to be interrupted several times contradicts the expectation of a smooth trip, which tends to be taken for granted in Western experience of travel.

I provide this report of my journey at the beginning of this contribution to illustrate central aspects of the relationships between travel and roadside communities. In a desert space, solidarity is an all-important principle. Thinking in terms of solidarity may have guided both the Teda driver who agreed to pick me up as well as the hospitable soldiers. I will return to address this topic later. A further point of this narrative is relevant too. As I already mentioned, in order to 'master' the desert, staying in one place is not useful and this explains why I had to leave Turayt and continue eastwards – despite the hospitality of

4 Although the women had paid a higher fee for the cabin, they were only 'loaded' on the floor of the pick-up. Other vehicles carried twenty-seven male migrants, while there were less than ten women and girls on our car, which made the passage more comfortable. I presume that this was another reason for the soldiers to propose this driver to me. During the journey, we picked up several male migrants from other overloaded Hilux.

the soldiers, the luggage left behind and the uncomfortable conditions of a migrant transport.

Through 'Tuareg Country': From Agadez to Assheggur

Caravan trade between Agadez and the Aïr region in the west and the Kawar, Bilma and Fachi in the east was and still is controlled by the Tuareg from the Kel Ewey and Kel Gress. They bring millet which they sell or exchange for salt from local production and dates from gardens and oases. The Tuareg monopolised the salt trade since 1759 when they defeated the Bornu Empire and took control of the area (Fuchs 1989: 118–120).

Caravans to and from Agadez and Bilma, called *taghlamt*, could range from enormous enterprises with several thousand camels to smaller groups with less than ten animals. Still, in 1946/47, one caravan of 25,000 camels came to the east bringing 350 tons of millet in exchange for 1,500 tons of salt and 1,000 tons of dates (Chapelle 1982: 112). Today, however, this business has diminished considerably. Prior to the French taking power in the region at the turn of the 20th century and ensuring the security of the caravans by military protection, the *taghlamt* repeatedly became the target of raids carried out by Weled Sliman Arabs and by the Tubu. In 1899, for example, 130 Arabs coming from the Fezzan and 400 Tubu attacked the caravan at the salines of Bilma, killing seventy Tuareg, injuring many others and making them lose about 7,000 camels (Chapelle 1982: 112).

Whereas Kel Ewey, who live in the Aïr Mountains came directly to Bilma on the road I now return to, Kel Gress had to take a more southern track leading first to Fachi. This track still exists but it is difficult to navigate because of the high dunes before reaching Fachi. In order to control the salt-trade, the Kel Ewey seemed to have guarded 'their' northern road avidly and to have allowed no caravans (with some exceptions from the Kel Gress) to travel from Bilma to the southern Hausa markets via the Aïr (Spittler 1984). Because the salt transport was the business of the Tuareg and a question of spatial dominance, the Teda, of course, did not participate in these caravans (Fuchs 1983: 126–132). Furthermore, toponyms alongside the road which the Teda still use, such as 'Turayt' or 'Assheggur', are Tamasheq names and should be seen as a testimony of former Tuareg dominance in this region. 'Turayt' may relate to *tareyt* ('way') or designate a rocky escarpment – the place is characterised by a small mountainous massif. 'Assh-eggur' may name a place where jackals (*eggur*) drink (*esshu*) (oral communication by Mohamed Aghali-Zakara/Paris). Even the whole desert from Agadez to the Kawar is referred to as 'Tenere' in Tamasheq, which means 'desert' and generally denotes an empty wasteland, whilst in Tedaga there is no proper name for it.

Depending upon different perspectives, the Tenere today is either still considered as 'Tuareg country' or as 'no man's land'. Either way, it is feared by those who are not accustomed to crossing it – stories of robbed merchants and of hidden bandits are numerous and frequently recounted. To these, one can further add geopolitical information on terrorist activities in the Sahara. Though difficult to verify (Keenan 2009, 2013), it is not without reason, however, that Nigerien army provides a weekly military convoy which the large majority of travellers use. When talking about desert travel, the Teda contrast the Tenere with other deserts. They claim that the crossing of the Tenere can be dangerous, whereas travelling in the Kawar, the Djado and even the Fezzan does not present major risks. Such a view may also be nourished by the ancient rivalry between Tubu and Tuareg – as both carried out raids against each other prior to the arrival of the colonial forces and both contest each other's claims over the Kawar (Chapelle 1982: 52–53, 112, 332–333; Spittler 1984).[5]

On our own track, the roadside community of bandits held only an imagined yet decisive presence. They were feared for their rapid incursions, which Klute (2009a) describes as part of a strategy of 'small wars' in the desert. Such potential incursions cause Teda drivers to respond in their own way, by using fast vehicles in combination with the military protection of a convoy.

The entire journey from Agadez to Sabha with a new but overloaded Toyota Hilux takes two or three days, with the part from Agadez to Assheggur (about 420 km) less than twenty-four hours. Most of the Teda drivers join the Agadez-Dirkou convoy protected by the Nigerien military, which departs every Monday evening and returns on Thursdays. After approximately seventy-five kilometres, the convoy reaches Turayt, which is the last military post before the desert. At this point, each participating vehicle and driver is supposed to be registered and passengers are meant to show their identity cards. The vehicles

5 Narratives of such battles between the two peoples are still alive. The most prominent among the Teda seems to be the battle of Ferwan, a place near Séguedine, where Teda defeated Tuareg who had just carried out a raid on the Tibesti. Many Tuaregs were killed. After this battle, Tubu and Tuareg concluded a treaty of non-aggression (oral communication by an interlocutor of the Kawar village of Beza). According to Chapelle who is referring to the *Chronicles of the Aïr*, the Tuareg defeated the Tubu in 1777. Since that time, Kanuri people of the Kawar, who formerly marked their livestock by brands of the Teda Gunna ('Gounda') and Tomara ('Tomagra'), had to use brands of the Kel Ewey. Teda raids were usually small enterprises of four to five, rarely more than ten participants, and could reach regions, which were about 1,400 km away from the Tibesti (Chapelle 1982: 111, 333–335).

then reach Kori-n-Kantana late in the evening or even in the night where they rest for some hours.

Early the next morning, the convoy leaves Kori-n-Kantana and very soon a Hamada-like stony desert begins, with the Aïr Mountains to the north. After approximately 290 km from Agadez, the desert becomes sandy. Drivers begin by crossing several dunes and later, their route continues parallel to the alignment of the dunes, which now flank the corridor for the vehicles.

Around noon, our vehicle (one of the first cars) arrived at *Puits de l'espoir*, a cemented well covered by a kind of hangar with plenty of waste in its surroundings. By early afternoon, we reached the well of Assheggur where we took a rest. From here, most migrant transports go straight to the north, thus bypassing Dirkou and its numerous controls, such as the military, the gendarmerie and customs.

On our return, our convoy left Dirkou Thursday evening and we spent the night at the well of Assheggur. Early the following morning, we continued to reach the stony Hamada by early afternoon. We had a rest at Kori-n-Kantana, passed Turayt late in the afternoon and arrived at Agadez that same evening.

The bivouacs at Kori-n-Kantana on the way to Dirkou and at Assheggur on the way back to Agadez are assembly points where vehicles gather when authorities are still busy with controls in Turayt or Dirkou. After having controlled the leaving vehicles during the day and the evening of departure, the soldiers join them late in the night at the bivouac. The following morning, the convoy leaves, protected by an advance party and by a rear-guard.

Former Nomads Become Drivers

The Teda have always been a heterogeneous group with variety of different lifestyles. In this contribution, I neither want to reproduce a speculative opposition between 'modern driver' and 'traditional nomad' nor to romanticise a nomadic group. More recently during the Chadian civil war (1965–79), the Teda became globally known as 'modern' actors in an international crisis (see Buijtenhuijs 1978, 1987) rather than 'traditional' pastoral people. Today, the business of the migrant-transporting drivers has an important geopolitical impact. Nevertheless, 'nomadic past' and current geopolitical importance seem to be closely linked: due to a history (and present) as a highly mobile people, most Teda know how to make roads, perhaps better than anybody else does. Taking over the ancient Tuareg caravan trail is obviously based on such long-standing experiences of road making adapted to a modern context. The use of fast and new Toyota Hilux pick-ups instead of camels allows for not only competing successfully over a 'foreign' road, but also involves changes in orientation and spatial perception.

The Teda are very mobile, most have relatives across the borderlands of Niger, Chad and Mali, and they therefore feel foreign nowhere: 'Here in the desert, I feel at home, because I have some family members wherever I will go.'[6] Moreover, the Teda always appear to be ready for departure. Jean Chapelle – a former officer of the French camel infantry and prefect of Chad and who was involved in the recruitment of Tubu for the famous camel-troops (*goumiers*) – emphasised their knowledge in orientation, their ability to travel under rough conditions and their willingness to leave at any given moment (Chapelle 1982: 173–175, 335, 339).

In order to cope with the exigencies of modernity and to compete for mastering the Sahara, the Teda often buy and use cars. Many drivers own their vehicle – a rich kinsman from Libya may have given a car to a young driver, and the latter, thanks to his remunerative activities, may have rapidly reimbursed his debt. The transport of migrants offers a highly profitable business: a total of twenty-seven passengers can be 'loaded' onto a Toyota pick-up, and each of them pays at least FCFA120,000.[7] On the way back from Libya to Agadez, the passengers – returning labour migrants from Libya – are less numerous, but to compensate, the Teda load more 'smuggled' goods. Drivers told me that they can buy a new Toyota Hilux after three successful trips, and many of the Hilux in the convoy were bought new in Libya the year preceding my research.[8] As the Teda's Hilux are not old imports from Europe (as in other parts of Africa) but new ones bought in Libya (one driver proudly showed me the traits in the design of his vehicle, which proved that it was a very recent model), creative adaptations may be minor. Nevertheless, the drivers carry them out.[9]

---

6 I am citing here the chief of a driver's cooperative, but the same idea was expressed by other interlocutors too. The chief further explained 'Libya is for us, Chad is for us, Niger is for us. If a Tubu has to go to Chad, he will again take a wife there. This is in order to have relations. Still another thing: If they [the Tubu] have to leave for a far place, they will not marry kin. They marry a stranger. In this way, they multiply their relations. The house of your kinsman is your house.'

7 Euro is about FCFA656.

8 The cost of a Hilux in Germany was Eur22.134 (July 2014), according to the Toyota website (www.toyota.de) consulted the same month. A more basic and thus cheaper version may be available in Libya. In Nairobi, the cost of a Hilux was about £10,000 (see Carrier 2009). In 2011 and 2012, many Hilux seem to have also been removed from their Libyan owners.

9 For example, in order to 'load' twenty-seven people and to avoid that they fall down when travelling with the speed of 100–120 km/h on rough terrain, the drivers fix wooden sticks around the load-floor for passengers to hold on to. Or tyres are deflated when entering sandy areas in order to avoid getting stuck.

FIGURE 9.2 *Convoy tracks leading eastwards towards Puits de l'Espoir*
SOURCE: T. MUSCH, 2014

When following the dialectics of appropriation (Beck 2002, 2009) one could suppose that owning vehicles makes their owners themselves undergo change. Disregarding any romantic notions, one has to agree that riding a camel and driving a car is not the same thing and in particular, such change seems to have an impact on the perception of the landscape. When speaking with Teda drivers about orientation in space, I was told that the nature of orientating depends on the means of locomotion: for example, landmarks relevant for orientation when travelling by camel or by foot are different from landmarks used when travelling by car. This fact may have two reasons. Firstly, a car cannot go everywhere a camel or man travels, but has to be driven on more or less appropriate terrain, which leads to detours from 'traditional' ways. Secondly, the perception of landscape during a rapid ride by car is different than during slower travel, and when other 'landmarks' are used for orientation (see also Spittler 1998: 40–51; Laurier 2013).

'Traditionally' the Teda use the position of the sun, stars or landmarks such as mountains, rocks or perhaps some lonesome tree for orientation. 'Finer' landmarks also play an important role such as the alignment of dunes, changes

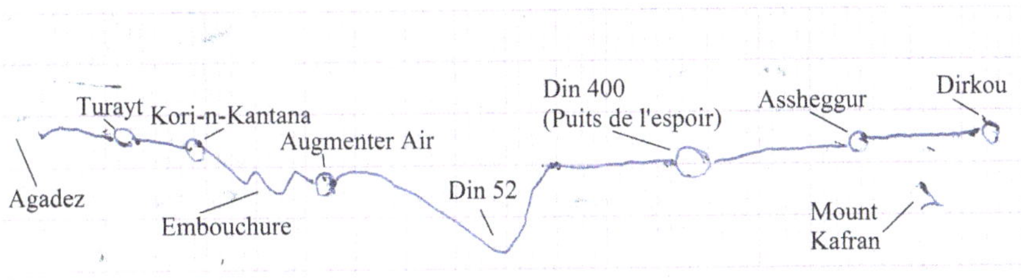

FIGURE 9.3 *Mental map Dirkou-Agadez drawn by a Teda driver*
SOURCE: MOUSSA KALIKORE, 2014

in colour or granularity of sand, the emergence of depressions with white lime or black-reddish basalt or the patterns left by the wind on the ground. The latter seem particularly important when there is no more orientation. A Teda narrated how he and his driver got lost once. He explained that it was a new moon and that the clouds covered all the stars. The driver stopped, left the car and regained his orientation only with the help of the ripples in the sand, which indicated the prevailing wind direction as well as its intensity.

Most of these signs are not used when driving a car at high speed and without stopping, as they tend not to be recognised precisely. This is why drivers increasingly use mobile phones with integrated GPS, bought in Libya.[10] They are thus able to document important stages of their journey and if necessary, the device guides them from one point to another.

The names of places that the drivers use in the GPS are often written in localised French.[11] Several of them are not 'traditional' toponyms of the caravan route but most likely to have been recently invented (e.g. *Augmenter air, Embouchure*). This suggests that a new way of knowing the road has been established by young Teda drivers who have developed a new way of documenting and perceiving the road. Current practices such as borrowing mobile phones, selling used ones among friends and the sometimes collective use of the expensive Thuraya satellite phones (along with integrated GPS) contribute to the diffusion of such knowledge among the travel community.

The author of the mental map reproduced above (see Fig. 9.3) was a thirty-five-year-old driver who drew it for me whilst holding his mobile phones and

10 Such mobile phones are available in Libya for the equivalent of about FCFA120,000 (new).

11 Throughout the text, I tried to write Tamasheq names in a way close to Tamasheq transliteration in order to emphasise their Tuareg origin. The drivers note them in a different, slightly francized but not official way. Examples: Assheggur = Achougour; Turayt = Tarewete. *Din 400* written in Franco-French means *Dune 400*.

consulting its GPS-data. In good weather conditions it is not necessary to drive from the Kawar to Agadez or back with the help of a GPS and the drivers (even the first ones) only have to follow the numerous tracks of precedent vehicles or convoys in order to arrive without fail at the places of the road marked on the map. With bad weather conditions or when a driver becomes inattentive and loses the main road, the GPS turns into an important tool for finding the way.

The use of the GPS attests to a different mode of spatial orientation, which differs from orientation when walking or riding a camel. Several of its waypoints are not really landmarks that would be helpful for orientation, but rather places that are important for safe and secured motorised travel. Thus, the points called *Din 400* (Puits de l'espoir), *Din 52* or *Augmenter air* show the presence of dunes to be surmounted or circumnavigated or the beginning of a rocky terrain, which needs well inflated tyres. The post of Turayt, the valley Kori-n-Kantana and the well of Assheggur are not landmarks for visual orientation, but places to pass checkpoints or to rest and to stay in contact with other drivers (see below). A final point called *Embouchure* marks the dangerous narrowing of the track where ambushes may be possible. Only a spike in the east symbolising mount Kafran can be fully considered as a landmark that is important for orientation. As vehicles arrive here from many directions (Séguedine, Dirkou, Bilma, Fachi), the alignment of wheel tracks may be somewhat confusing, and a prominent landmark which is even visible when driving at high speed can be useful.

When I observed the driver drawing the map and when I listened to narratives of the road from other Teda, I noticed a symbolically important detail: they always started in the Kawar and then came to Agadez. However, in terms of economy, the opposite would be more logical as the majority of migrants come from Agadez, which is also the region where the transport business started. Moreover, for centuries during which the Tuareg travelled this road, the starting point of the *taghlamt* was in the west and not in the east. Yet the orientation of my Teda interlocutors is inversed. At least two features of the map underline this orientation from east to west:

1. The map has a starting point marked with a round circle (Dirkou), but not a precise end. Agadez is not even presented on the map, only a line symbolises the way to it from Turayt. 2. An important place of the journey is named *Augmenter air*. That means that vehicles coming from the east entering the Hamada and leaving the sandy desert have to inflate their tyres. When coming from Agadez, tyres are deflated in order to prepare the vehicle for sandy areas, which start here. The fact that the place is called 'Add air' and not 'Diminish air' corresponds with discourses of Teda drivers who are narrating the road from the east to the west and not the other way round.

On the road from Agadez to Dirkou, just some kilometres before *Din 400*, our driver took a wrong dune corridor, which narrowed more and more and then led to a dead end. The driver calmed down some puzzled passenger by saying not to worry as he allegedly knew the track 'since his childhood'. Conversations later revealed that this information was not very accurate. But it emphasises the claim that 'since always' the road 'belongs' to the Teda.

Yet, still more interesting than this claim of ownership was that the driver did not return immediately, but stopped, left the cabin and walked around in the sand for some time. I do not know whether he searched for some clues. Perhaps he tried to regain concentration by a 'pause', similar to a procedure that Chapelle describes for the Teda who have lost orientation (1982: 175). In this case, the traveller stops, closes his eyes, turns several times around himself, lays down and waits until the dizziness vanishes. Then he gets up and looks at the stars and the landscape 'as if this was something new'. He thus recaptures orientation and continues.

Road Making by Interaction

In the morning at Assheggur or Kori-n-Kantana, just as the sun rises, dozens of Toyota Hilux are starting their travel. This is an impressive spectacle, not least because when arriving here at night, one does not really notice how numerous they are. Most of these cars have Teda drivers, and all the drivers seem to know each other to a certain degree. How do these Teda whom Chapelle describes as 'deeply individualistic' and 'unsociable' (1982: 339–340) interact in order to compete collectively for the 'foreign' road?

I have already mentioned that when travelling in the desert, solidarity is very important. Despite Chapelle's characterisation, solidarity serves drivers as a principle. Cautious Teda drivers may form loose groups – as they did in the time of raids – of three to five vehicles and travel within eyeshot of each other or wait for each other at several points of the road. More individualistic natured drivers who travel alone can however count on the solidarity of their colleagues and, in case of breakdowns or other incidents, receive help from others. In any case, every driver is waited for at the final destination of the road (Agadez and Sabha) and sometimes even before, by co-drivers of the same convoy, by friends or relatives; and should he not arrive as expected, they claim 'we all will search for him'.

Thus, the bivouacs at Assheggur or Kori-n-Kantana, which serve authorities for better organising and controlling the convoy, serve the drivers for another kind of 'control': every one of them can show others that he is in the convoy

FIGURE 9.4 *Bivouac at Assheggur*
SOURCE: T. MUSCH, 2014

and vice-versa. Once the presence of a driver is ascertained, his colleagues will be able to locate him in the case of his non-arrival and, later on, to search for him if necessary.

Solidarity also leads to the sharing of instruments. When we arrived at the Hamada from Dirkou, not every driver had brought an air pump and those who carried one in their car lent it to others. The same is true for the Thuraya satellite phone. Not all the drivers were equipped with this expensive instrument, but should one be in need of it, he will be able to find somebody in the convoy who will lend him his.

When speaking about travel communities, one must not forget the migrants. They are the principal motivation for engaging in the transport-business and the source of wealth for the drivers. The routes formerly taken by migrants which lead to the Canary Islands or the Mediterranean Sea via Mauritania, Algeria, Mali and Morocco are now difficult or even impossible to use because they are highly monitored or closed due to warfare (see also Choplin and Lombard 2013; Kastner 2014: 53–58). Libya, despite its current political disorder and the quasi absence of state power in its south, continues to remain an option for travelling to the north for West African migrants.

During the journey, passengers from Ghana, Nigeria and Burkina Faso told me that they could easily reach Agadez by themselves as several bus companies ply the entire region. Coming, for example, from Accra is neither expensive nor logistically difficult. The problem lies in crossing the Sahara, which explains why brokers – who probably have close links to an expanding community of Teda merchants in the town[12] – search and try to pick up migrants already from the bus stations in Agadez.

In Agadez, the migrants' agency in the making of the road ends thus abruptly. From here, they are assigned a passive role as they are merely 'transported' and sometimes, when a vehicle is overloaded, they are exchanged like any other cargo between drivers. As the migrants are 'southerners', they fear the desert and its people and therefore seem to be accepting of everything – ultimately, they do not have any choice. During the journey, the most salient reaction I could sense among the migrants was an increasing desire to leave the desert and not to be left there (this can also happen).[13] Migrants are thus personifying, albeit in a passive way and without pretending to master space, the axiom that staying in the desert is futile, whereas crossing it rapidly is essential.

Absent from the road in physical terms but present virtually are the Tuareg. The only place where I felt the real presence of Tuareg was Turayt. When I arrived during the night, the bleating of goats announced the existence of some camps in the neighbourhood, and, on the return, in the afternoon, some Tuareg women were begging for money. Certainly, Tuareg drivers would also like to engage with the remunerative business of transporting migrants. Yet, they are in a disadvantaged position concerning the appropriation of roads: whereas Teda are managing to take over a 'foreign' road – the ancient caravan track from Agadez to Bilma – the Tuareg would simply get lost when adventuring

12 A broker who can 'fill' a vehicle (twenty-seven migrants) earns the fee equivalent to one or two passengers.

13 Returning to Agadez from Dirkou, we came across a group of about twenty migrants from Burkina Faso on a dune. Their driver had told them that his vehicle may soon break down and asked them to leave and to wait. Allegedly, he pretended that the town was not far and that he would come back to pick them up. This had occurred during the convoy to Dirkou (probably on a Tuesday) and we found them Friday morning. The driver had left them in the desert with just some water but without food or blankets for three nights. In order to avoid such situations, migrants accept everything. Upon our return, when one of our cars already broke down near Turayt, migrants feared to be left there and more or less voluntarily paid \$50 to a driver of another vehicle to continue their journey. Nevertheless, as a rule, no money is to be paid when a vehicle breaks down, and the passenger continues with another one. (For more information on deprivations of migrants during desert travels, see also Triulzi 2013).

into the Djado and the southern or central Fezzan. Also the Kawar region today seems not to be very familiar to Tuareg. In these places, where the Teda are 'at home', the rather 'foreign' Tuareg cannot count on a 'third' power, such as the military convoys that ensure protection for travellers in the Tenere.

Tuareg, who formerly extended their power to the Kawar, are not involved in the transportation of migrants on this road, and only some probably centuries-old names of places attest to their former sovereignty over this space. In the travellers' imaginations, the Tuareg were nevertheless present and 'shaped' the road: the danger which menaces travellers through the Tenere and which makes the military convoy necessary was implicitly (and sometimes explicitly) attributed to Tuareg. The awareness of a history of raids and counter-raids as well as the supposed regional 'insecurity' contributed to the feeling of danger among travellers.

In order to protect travellers from imagined and real dangers, Niger's army provides a weekly convoy. The convoy is not only used by Hilux-driving Teda, but also by lorry drivers, state service vehicles, NGOs as well as by private travellers. One may question how Teda drivers – who proudly described themselves to me as the (sole) 'masters of the Sahara' – may then subdue to the military. Although the Teda can become skeptical when they encounter a 'foreign' power in 'their' space, due to the imagined threat of the Tuareg on the traditionally foreign road between Agadez and Assheggur, they accept the support and participate in the convoy. Their participation can be realised in two different ways. Either the driver is on friendly terms with the soldiers at Turayt, in which case he will usually be able to pass their controls and continue,[14] otherwise they join the convoy only after the post, which is usually the case. Most of the Teda drivers join the convoy late at night at the bivouac of Kori-n-Kantana in order to avoid controls by the gendarmerie around Turayt. Upon returning, many join the convoy only at Assheggur, after the controls of Dirkou.

How do the 'proud' and 'freedom-loving Teda'[15] react when obliged to use a convoy and to be subordinate to state military power, thus recognising the

---

14 Here, I refer to the example of my driver on the way to Assheggur. He was known and 'recommended' to me as a 'good' driver.

15 Young Teda always presented themselves as 'freedom-loving', 'independent', 'adventurous' and 'audacious'. See also Chapelle's characterisation of the Tubu: 'They seem at a first glance like a people of pariahs of the desert, however they are a free people who knew, during centuries, how to save the independence of their tents, to live as they liked to do, and to be feared and respected by their neighbours' (1982: 1). Teda drivers emphasised in particular the fact that 'nobody loves' them (they meant: no government loves them) because they engage in 'forbidden things', for example transport considered as 'smuggling'.

supremacy of the latter? Drivers may cope with such a 'submission' in two different but complementary ways. Firstly, they may react pragmatically. The 'hostility' of the desert makes solidarity necessary and blurs lines of antagonisms, such as between 'smuggler' and 'military'. Not only do drivers accept pragmatically the protection afforded by the military, but also the military accepts to protect drivers who may, in another context, be considered as 'smugglers'. Secondly, and more crucially, a kind of 'creative modification' of the convoy may help the Teda drivers to avoid submission as they are only using the convoy for the stony Hamada in the south of the Aïr Mountains where the threat seems indeed to be present. The further they enter the sandy areas, the more they feel familiar and free. On our way eastwards, just after the Hamada, the Hilux started to overtake the advance party of the convoy, head towards the Kawar at high speed and turning the first military vehicles into a kind of rearguard for their own security or even into a kind of lure supposed to distract bandit raids.

The following narrative once more illustrates my point regarding the pragmatism of desert travellers. On our way towards the Kawar, we took a break at the well of Assheggur, along with other Teda drivers. Two vehicles of customs officers travelling in the same convoy also arrived. They were on their way to relieve their colleagues in Dirkou and they camped out near us and eventually started making conversation with us. If the Teda drivers and the customs officers would have met in the borderlands of Libya and Chad, it would have probably been a less amicable encounter. But here, as members of the same convoy and as fellow travellers, they felt obliged to conform to a kind of pragmatic solidarity.

Assheggur is the last place on the road eastwards where such solidarity works. Further on, in the areas that are closer to the border of Chad and Libya, the travel communities will behave differently. That is where the military patrols and customs officers are searching for 'smugglers'. For the drivers, there is no more need for military protection as the space 'belongs' to them. During the crossing of the Tenere, on the road that traditionally 'belongs' to the Tuareg, both the military and the Teda drivers may feel like 'foreigners', whereas in the Kawar and the Djado, only the soldiers feel as 'foreigners', and the Teda feel 'at home'.

After our rest at the well of Assheggur, my driver told me that he could not continue to Dirkou but needed to go straight to the northeast, towards Sabha[16] as the many controls around Dirkou would be too expensive to pass and

16 From Assheggur, the drivers continue to the Kawar village of Séguedine, reach Madama in the far north, cross Tummo, and then head for Sabha.

perhaps not all vehicles would even pass these checkpoints. The space of pragmatic solidarity was now about to finish, and the driver's strategy turned into hiding and escaping.[17]

Conclusion

Tuareg influence on the road between Agadez and the Kawar has diminished considerably in the last decades. Formerly, they fiercely defended their caravan route against competitors on the regional salt market. Teda drivers, who used to be rivals of the Tuareg, have increasingly taken over the desert routes with the transport of migrants. As I have shown, they do so by means of their high mobility, which enables them to appear and disappear in different places. Their mobility is not only result of the rapidity of the Hilux and the Teda's readiness to travel, but also of the recent and 'illegal' character of their business. This allows them to travel relatively unrestricted by rules, except for those connected to solidarity. For example, Teda drivers can make 'use' of the security provided by military convoys whenever needed and without having to follow the state's rules. This kind of mobility particularly pertains to those spaces in which Teda have lived and travelled already since centuries.

The changes subsequent to the Teda's appropriation of the caravan route have affected not only its ownership. The Teda, as a travel community, have also undergone changes. Their appropriation of the road, along with newly acquired means of transport (mainly Hilux pick-ups and GPS-devices) engendered new perceptions of space. This becomes evident, for example, through the modified naming of places alongside the road according to the priorities of secured travel of the road's new masters. The fact that the road was formerly imagined as leading from west to east and is now 'inversed', starting in the Kawar and running to Agadez, is another example of a different perception and of making of the road.

The long-standing practice and experiences of travel as an outcome of a culture of high mobility guides the Teda in their road making. At the same time, such experiences have to be adapted anew to the 'modern' conditions of desert travel. The referencing of 'new' landmarks, for instance, does not only express

17 Due to the petrol-driven engines of their Hilux, the Teda can drive faster than the soldiers who use diesel-engines. Furthermore, thanks to the excellent topographic knowledge in their 'homeland', the Teda can hide and even travel at night. One driver explained: 'Sometimes, we are driving along a dune corridor and the military drives in the neighbouring one without being aware of us.'

new perceptions of space, which reflect the new ownership of the road, but it is also a result of new devices for orientation.

There is a relationship between the way in which people perceive space and how they interact with each other on the road. The more a road is imagined as dangerous, the more important pragmatism and solidarity become among groups of travellers. Imaginations of danger do not only arise from the hostile environment of a desert road, but also from other travel or roadside communities, in this case, the potential threat of Tuareg bandits.

Social relations between various road and roadside communities decisively shape the road and the choice of road making strategies, such as hiding and escaping, searching protection and thereby accepting control, or the opposite: dodging protection and thereby evading control. All these strategies require mobility and staying, on the contrary, is not a strategy in the desert. Only by moving can one experience and appropriate space (Musch 2013). This is what the statement 'we are free, we are the masters of the Sahara' means. Always being free to leave a place for another one is not only a condition for coping with the desert, but also seems to be the condition for claiming ownership in the Sahara.

References

Beck, Kurt. 2002. 'Die Aneignung der Maschine.' In *New Heimat,* edited by Karl-Heinz Kohl and Nicholas Schaffhausen, 66–77. New York: Lukas and Sternberg.

Beck, Kurt. 2009. 'The Art of Truck Modding on the Nile (Sudan): An Attempt to Trace Creativity.' In *The Speed of Change: Motor Vehicles and People in Africa, 1890–2000,* edited by Jan-Bart Gewald, Sabine Luning and Klaas van Walraven, 151–173. Leiden: Brill.

Beck, Kurt. 2013. 'Roadside Comforts: Truck Stops on the Forty Days Road in Western Sudan.' *Africa* 83 (3): 426–445.

Buijtenhuijs, Robert. 1978. *Le Frolinat et les révoltes populaires du Tchad. 1965–1976.* The Hague: Mouton.

Buijtenhuijs, Robert. 1987. *Le Frolinat et les guerres civiles du Tchad (1977–1984).* Paris: Karthala.

Carrier, Neil. 2009. 'The Hilux and the "Body Thrower": *Khat* Transporters in Kenya.' In *The Speed of Change: Motor Vehicles and People in Africa, 1890–2000,* edited by Jan-Bart Gewald, Sabine Luning and Klaas van Walraven, 175–190. Leiden: Brill.

Chapelle, Jean. 1982 (1958). *Nomades noirs du Sahara: Les Toubous.* Paris: L'Harmattan.

Choplin, Armelle and Jérôme Lombard. 2013. 'Stranded in Mauritania: Sub-Saharan Migrants in Post Transit.' In *Long Journeys: African Migrants on the Road,* edited by Alessandro Triulzi and Robert L. McKenzie, 69–94. Leiden: Brill.

Fuchs, Peter. 1983. *Das Brot der Wüste: Sozio-Ökonomie der Sahara-Kanuri von Fachi.* Wiesbaden: Franz Steiner.

Fuchs, Peter. 1989. *Fachi: Sahara Stadt der Kanuri.* Wiesbaden: Franz Steiner.

Hahn, Hans-Peter. 2004. 'Die Aneignung des Fahrrads.' In *Blick nach vorn: Festgabe für Gerd Spittler zum 65. Geburtstag*, edited by Kurt Beck, Til Förster and Hans-Peter Hahn, 264–280. Köln: Rüdiger Köppe.

Kastner, Kristin. 2014. *Zwischen Suffering und Styling. Die lange Reise nigerianischer Migrantinnen nach Europa.* Münster: LIT.

Keenan, Jeremy. 2009. *The Dark Sahara: America's War on Terror in Africa.* London: Pluto Press.

Keenan, Jeremy. 2013. *The Dying Sahara: US Imperialism and Terror in Africa.* London: Pluto Press.

Klute, Georg. 2009a. 'Kleinkrieg und Raum.' In *Begegnungen und Auseinandersetzungen. Festschrift für Trutz von Trotha*, edited by Katharina Inhetveen and Georg Klute, 283–303. Köln: Rüdiger Köppe Verlag.

Klute, Georg. 2009b. 'Modern Chariots: Speed and Mobility in Contemporary "Small" Wars in the Sahara.' In *The Speed of Change: Motor Vehicles and People in Africa, 1890–2000*, edited by Jan-Bart Gewald, Sabine Luning and Klaas van Walraven, 191–211. Leiden: Brill.

Laurier, Eric. 2013. 'Driving: Pre-Cognition and Driving.' In *Geographies of Mobilities: Practices, Spaces, Subjects*, edited by Tim Cresswell and Peter Merriman, 69–82. Farnham: Ashgate.

Lynch, Kevin. 1976. *Managing the Sense of a Region.* Cambridge: MIT Press.

Musch, Tilman. 2013. 'Territoriality through Migration: Cases among the Tubu Teda Guna (Niger).' *Nomadic Peoples* 17 (2): 68–81.

Scheele, Judith. 2012. *Smugglers and Saints of the Sahara: Regional Connectivity in the Twentieth Century.* Cambridge: Cambridge University Press.

Spittler, Gerd. 1984. 'Karawanenhandel und Kamelrazzia bei den Kel Ewey: Die Kontrolle des Salz- und Hirsehandels zwischen Air, Bilma und Kano (1850–1900).' *Paideuma* 30: 139–160.

Spittler, Gerd. 1998. *Hirtenarbeit: Die Welt der Kamelhirten und Ziegenhirtinnen von Timia.* Köln: Rüdiger Köppe.

Triulzi, Alessandro. 2013 '"Like a Plate of Spaghetti": Migrant Narratives from the Libya-Lampedusa Track.' In *Long Journeys: African Migrants on the Road,* edited by Alessandro Triulzi and Robert L. McKenzie, 213–232. Leiden: Brill.

CHAPTER 10

Technological Dramas on the Road

The 'Artery of the North Highway' in the Sudan

Kurt Beck

Africa's economic renaissance in the decade after 2000 has been accompanied by large-scale investment in infrastructure, most notably in telecommunications, energy and roads. The revived Trans-African Highway Program, which aims at the completion of a highway network with a total length of nearly 60,000 km, is a visible witness to this boom (SWECO 2003). The Sudan, to which I refer in this chapter, has dramatically extended its paved overland roads network, doubling its total length in just six years between 2004 and 2010. As a consequence of this changing materiality of the road, the overall regime of automobility in the Sudan is rapidly changing.

My main interest in this contribution is to examine what happens when a region emerges from 'roadlessness' by the construction of a highway. There certainly exist cases of regions that have emerged from total roadlessness – but given the long history of automobility nearly everywhere in the world, today the construction of an asphalted road usually means the replacement of previously existing roads, whether gravelled roads, unsurfaced roads or simple tracks. In such cases, the road as well as the motor vehicles that travel it are not radical innovations, as there are already established practices of motorised travel. Nevertheless, as with changing material technology in general, the new materiality of the asphalted road allows for new practices of use to be developed, for new meanings to be attached to it and for different material technologies to be operated on and along it, in short, for new orders of the road to evolve.

My narrative draws on the case of the highway called *ash-Shiriyân ash-shimâl* (Ar. Artery of the North), which was completed as a toll road in the early 2000s and which connects Omdurman and the Sudanese capital Khartoum with the northern parts of the Sudan. *Ash-Shiriyân ash-shimâl* roughly follows the large *wâdî* called Wadi Muqaddam. It is part of the Trans-African Highway network and replaces an older unsurfaced road through the Wadi Muqaddam area.[1]

1 My first acquaintance with travelling Sudanese style goes back to 1975 when I visited the Sudan as a student. In the early 1980s it developed into a tiring, but otherwise completely normal way of reaching a field site for my doctoral research with North Kordofan nomads in

 | DOI 10.1163/9789004339040_011

Despite its huge construction programme since the turn of the century and as in African countries generally, paved roads in the Sudan represent only a fraction of all roads. I argue that road travel is thus situated in two distinct but interlocking technological regimes. In my North Sudanese example, I roughly distinguish between two different orders or regimes of the road. First, an older regime of desert travel on unsurfaced roads, as exemplified by the old Wadi Muqaddam road. This regime also comprises the small mechanics' and vehicle body manufacturing workshops, that modify and adapt imported lorries for their use on Sudanese hinterland roads (Beck 2009; Hänsch 2009), as well as the communities of drivers, traders and workers who provide transport for 'roadless' Sudan (Wadelnour this volume). The second order represents the modern highway regime, for which the *Shiriyân ash-shimâl* is an example and which is related to the experience of the urban Sudan of the petrol boom after 2000 and the internationally connected import economy, the regulatory state and the shiny products of Sudan's GIAD motor manufacturing company.

The point I thus wish to make is that the highway is much more than simply a strip of asphalt. I conceive of it as part of a larger technological regime and as carrying the potential for the reordering of the political economy of a whole region. This is why I aim at a perspective on the road's order, which first includes both the road and the roadside, and which, secondly, allows for the development of an argument about technological materiality and its appropriation by users.

Technological materiality may come in two different guises that significantly differ in terms of the uses to which the technology can be put. The first is of the sealed kind, which cannot be opened by users without it being destroyed. Consequently, users may choose between adopting it wholesale or leaving it.

---

a place that took four days to reach by lorry on the so-called Forty Days Road. I have visited large tracts of the Wadi Muqaddam area during ethnographic research with these nomads in 1983, 1987 and 1988 coming from Kordofan by camel. In the early 1990s and later in 2006 and 2008 I have travelled the Wadi Muqaddam road and later the highway to another field site on the Nile. As my research was directed to other issues, I used the road as a naïve user. The idea of turning the road itself into an ethnographic site appeared only in 2008 while doing research for an ethnographic documentary in truck building workshops with Valerie Hänsch. We thought it would be interesting to study these trucks in the wild. The opportunity came in 2010 when I launched a project on the African road together with Gabriel Klaeger, Michael Stasik and Rami Wadelnour. To all of them I owe a debt of gratitude for inspiration. I also would like to thank the German Research Foundation (DFG) for their generous funding. I also wish to express my deep gratitude to the late Hajj Adam wad Jakdul for being my companion and camel-guide in 1987 and 1988, to my Rûbâb hosts along the highway in 2013 and in particular to Hammad al-Karis who drove me around in his deep *mahallî* car.

The second kind comes as a technology, which lends itself to unpacking the technological package and selectively putting its components to uses that might not have been intended in its original construction. To maintain the assemblage of material artefacts, practices and meanings stable, these technologies require implementation, which is simply another way of saying that they have to be controlled in their use to prevent users from unintended uses. A perfect example is provided by the imported lorries that are ingeniously adapted to the Sudanese environment in the workshops referred to above, and that have, in this process, been modified beyond the confines of orthodox car manufacturing. This is what I call appropriation, namely to tamper with the meanings, the materiality of a technology and the practices associated with it. The technological materiality of the highway is clearly of this second kind. It is open to all sorts of unauthorised and unforeseen uses, such as, for instance, driving cattle along it, turning it into a moving market (Klaeger 2012) or a playground for children.

In some ways, the highway renders the state a reality in an area previously only very lightly administered, as the state makes large efforts to closely monitor and control the highway in order to keep the technological assemblage stable. As such, the highway embankment marks a border, which the powers regulating the highway regime guard against the incursions of those who adhere to the older practices of the regime of the unimproved road. Friction regarding the legitimate uses of the highway is likely to arise at the very interlocking of the two road regimes, and this provides the stuff from which technological dramas on the road are made.

What I refer to as technological dramas[2] are in fact political dramas veiled in a rationality of technical norms, for they are about participation and expulsion. By the very expansion of the highway regime, practices of the older regime of the unimproved road do not simply become obsolete but their practitioners become reconstituted as a hazard to road safety and the road generally. They become the undesirables of the road, simply due to intergovernmental declarations of a technological zone in connection with the Trans-African Highway Program and the efforts of their government, which attempts to link up to these international standards by regulating the road regime accordingly.

A technological zone can be understood, according to Andrew Barry, as 'a space within which differences between technical practices, procedures and forms have been reduced, or common standards have been established' (2006:

2 Taking up a suggestion by Bryan Pfaffenberger (1992) who refers to societal contests about technology as technological dramas.

239). DIN (originally German Industrial Norms) or ISO (International Organisation for Standardisation) norms, for instance, play a critical part in measurement, connecting technologies and regulation, and they apply to construction and information technology as well as to health and services technologies. It is easy to see how a common set of standards ranging from manufacturing and environmental to educational, political and legal domains can unify a sphere of production, consumption and transportation. In the case of the Trans-African Highway Program, the adoption of standardised norms of the road throughout the road network is of considerable importance, particularly when it comes to cross-border connectivity, and intergovernmental committees busy themselves with working out recommendations for national governments to adopt. Historically, these kinds of standardisations have been achieved by the nation state, indeed, they have provided an important foundation of the nation state itself by reducing heterogeneity and diversity, as Eugene Weber (1976) convincingly demonstrates in the case of France. Anthropologist James Scott (1998) has drawn attention to these homogenising processes as techniques of legibility, ordering and control. Global and transnational ventures like the Trans-African Highway Program have expanded technological zones and have created new technological zones. Being within or outside the particular technological zone of the highway clearly matters, as conformity to standards can convey great advantages. At the same time, technological zoning – as the expansion of the highway regime in the case at hand – can, and usually does, exclude others from linking up and instead turns them into undesirables.

Barry, as well as Scott or Weber, caution us against taking the expansion of technological zones for granted. In France for instance, Weber refers to a law of 1839 that had turned the metric system into the only legal system of measurement. By the end of the century however, people still used the old measures – especially in areas that were slow in entering the national market – and they even refused to speak French instead of their own languages and dialects (1976: 30, 32). An enormous diversity of measures and systems of measurement continued to be used prior to World War One, Weber argues, until finally the forces of colonisation from Paris – military conscription, compulsory school attendance, markets, wage labour and the spread of manufacturing, roads and railways – succeeded in creating a unified technological zone and a nation. Eventually, the story of the standardisation of France became a success, at least from the perspective of the state. The success, however, did not occur by itself, and we can imagine the technological dramas involved in that process. What, then, can we expect from the Sudanese order of the road? And how do these attempts at technological zoning and regulating the road regime translate at the road-level?

The Making of the Wadi Muqaddam Road

Rather than following the large bend of the Nile, north-bound travellers from Omdurman take the direct road through the desert that reaches the river again after 304 km on the highway at a place called al-Multaqqa (Ar. junction), between Ed Debba and Korti. Veering right from al-Multaqqa, the road then follows the river upstream to Merowe and the Hamdab dam. Turning left, it leads downstream to Ed Debba, Dongola, Mahas and eventually to Lake Aswan and the Egyptian border.[3]

For long stretches, the *Shiriyân* highway follows the course of the Wadi Muqaddam, which originates in the plateau west of Omdurman and flows into the Nile near Korti, some 25 km to the east of al-Multaqqa. Wadi Muqaddam itself follows an elongated depression lined with mountain ranges to the west and to the east from where during good rainy seasons small watercourses discharge their waters into the main *wâdî*. Good rainy seasons, however are rare events in this extremely arid region, which is the reason why the *wâdî* is usually dry. Whenever the rains in its catchment area are heavy (such as in 1988, 1999, 2006 and 2013) the *wâdî* floods large basins of clayey soils turning them into shallow pools for weeks and months. It is these clay pans, to which the nomads inhabiting the area claim traditional rights of cultivation and which can yield impressive crops of sorghum. Agricultural pursuits are otherwise dependent on irrigation. The Wadi Muqaddam is clearly better suited for the highly mobile nomadic keeping of camels, sheep and goats which can be fed on the perennial grasses, trees and shrubs of the *wâdî* and the thin cover of annual grasses and herbs, which, even after light rains, springs up in the adjacent areas with sandy soils called *qaizân* (Ar. dunes).

The Wadi Muqaddam has long been used as a passage from the confluence of the White and the Blue Niles near the current cities of Omdurman/ Khartoum and the savannahs to the west of it, to the place where it again meets the bend of the Nile between the old towns of Korti and Ed Debba (Budge 1907: 372; Grant 1875/6: 357). The latter was a convenient transit station because it was the farthest point regularly reached by sailing boats from Dongola downstream before the Nile bent east, owing to the constant north wind during the winter months (West 1918: 35). The arrival of the date fleet from the north, for example, was an important annual event throughout the first half of the 20th century. From the bend of the river, camel caravans took their loads to Kordofan and Omdurman. With the advent of the private motor car in the outlying regions of the Sudan from the early 1930s, commercial motorised transport

3 See map 10.1 on page 251.

appeared sporadically on the old caravan route. The availability of relatively inexpensive light commercial vehicles that were designed for rough US farm roads and their equipment with low-pressure air tyres facilitated the use of motorised freight and passengers transport throughout the Sudan, and even made it competitive with camel transport (March 1948: 181–183). In the Wadi Muqaddam, convoys of lorries transporting dates from Dongola to Omdurman were noted by the administration after Sudan's economic restrictions during World War Two eased (Lebon 1955). In the recovery year of 1949, the Governor of the Northern Province estimated that 500 tons of dates and 150 tons of hides were transported annually, as well as a monthly number of passengers amounting to no less than 500 (Sudan Government 1955: 160). In the mid-1950s, one observer passed no fewer than 14 southbound vehicles laden with passengers and dates on his 26-hour journey to the north (Lebon 1955: 273). By the 1980s, lorry traffic flourished in the Wadi Muqaddam to the extent – to use a local indicator – that travellers on a moonless night were nearly always able to detect the gleam of the headlights of at least one lorry on the horizon. Until the end of the century, mainly two desert roads were in use, one from Omdurman to Korti at the mouth of the *wâdî* itself and one branching off west to Dongola and approaching the river near Ed Debba.

Like other roads in the region, the Wadi Muqaddam road emerged without much notice from the administration once an affordable material technology became available for private use. Competences for using the road partly derived from older practices and skills of desert travel grafted onto its mechanised form, and partly from practices developed in the early state and military regime of motorised travel.[4] Caravan men were sometimes employed as guides

4 As generally the case in Africa, the first motor vehicles in the Sudan arrived around the turn of the 20th century. The first car in the Sudan was used by the Governor General's office (Sudan Government 1904: 222). In these early years, the administration already started to experiment with steam driven and petroleum driven motor cars for transport and with road building (Sudan Government 1906: 78 and 1907: 34) which is an exhilarating story in itself. In these early years until the 1920s, the motor vehicle was put to administrative and military use and later was also appropriated by some high-ranking Sudanese. The Governor General's Report for 1928 noted an increase in the number of motor cars and lorries from 849 to 2118 from 1926, 1554 already being owned by the public (Sudan Government 1930: 53). It was, however, only in the 1930s that private commercial interests came to utilise the car on a larger scale in the outlying regions and only after the severe import restrictions and requisitions for military use of World War Two commercially driven automobiles seriously conquered the roads of the Wadi Muqaddam type. At the same time an infrastructure of private repair workshops spread on a larger scale. Automobility in these earlier years was evidently organised according to a colonial regime that was built around government and military use, government

and lorries used to travel in convoy. Where drivers found them suitable, they followed the tracks of the camel caravans, with subsequent lorries in turn following their tracks, and this is essentially how the road emerged. In fact, lorries created the road by travelling it. Lorry drivers used to say *al-lastik bisallih as-sikka,* which means that the wheel itself renders the road motorable. During the dry season, lorries kept to the flat ground in sight of the *wâdî* where they could develop some speed; after the rains, they took to the higher and more rocky ground thus avoiding to getting bogged down in the clay pans. With time, interlacing bundles of tracks developed, all in a roughly north–south direction and suitable for different purposes and types of vehicles, be it for heavily loaded lorries, for double wheeled rear axles, or for small pickup cars and later the occasional 4x4 saloon car whose narrow wheelbases did not match the wide tracks of the lorries.

The road through the Wadi Muqaddam thus belonged to the type of road once described by a Sudanese geographer as 'made *by* cars and not *for* them' (Khogali 1973: 157, original emphasis). During colonial times, the road was apparently never deemed important enough to deserve serious upgrading, much unlike the roads used for cotton extraction from the Nuba Mountains to the railhead at El Obeid or the roads connecting the larger towns in eastern or southern Sudan. Those latter roads, leading through areas with different terrain and soil and situated in regions with high rainfall (which in turn led to high mortality rates from fly among transport animals, making motorised transport the only feasible alternative to head porterage) received periodic maintenance by gravelling and other forms of surfacing, as well as the construction of culverts, bridges and fords (March 1948: 181; Martzsch 1939: 120ff; Morrice 1949; Sudan Government 1906: 78ff).

In the Wadi Muqaddam not much could possibly be gained by upgrading, short of asphalting the road, which was out of the question. For all purposes, the road could be considered an all-weather road, as rainfall was a rare event and thus negligible. Moreover, the northern parts of the Sudan were well served by an excellent railway network of more than 4,000 km that had been constructed right from the beginning of the colonial Anglo-Egyptian government, and by 1912 it already connected Egypt, Port Sudan and El Obeid with the central region around Khartoum (Hill 1965). As part of a complementary and sometimes competing regime in the overall regime of Sudan's mobility

workshops, imports on government budget, and some private use by motor enthusiasts and nabobs of society. In this earlier regime of automobility, the motor car was clearly like an exotic beast in a colonial zoo, carefully guarded by its own gamekeepers, but by the 1930s it started to run wild and spread into the open range.

(much like the regime of river transport), the railway from Karima to Khartoum transported much of the goods and passengers of the Northern Province, which made upgrading the road a low priority. After particularly heavy rains, the District Commissioner's office sent a gang of prison labourers to those areas with denser vegetation to clear the road. The intention was more to weed fire-lines and prevent grassfires than an effort to repair the road. The nomads in the Wadi Muqaddam were extremely lightly administered both before and after independence and the road received attention only in very rare cases, such as when a high official of the Sudanese government travelled the road, for instance to attend an important tribal gathering or to ceremonially open a remote school for nomads' children. On such occasions a grader would be dispatched to flatten the worst stretches and the schoolchildren used to erect small heaps of stones on the last kilometres, less for demarcating the road than for decorating it in a gesture to honour the guest.

In the mid-1960s, short-lived plans were made for the implementation of a network of asphalted roads throughout the northern Sudan, including a highway following more or less the old unpaved route through the Wadi Muqaddam. Between 1965 and 1967 a team from Arizona State University contracted by Lockheed Aircraft International drew up a detailed scheme for road construction for the northern Sudan (Betz and Bauman 2007). This was part of a US Agency for International Development contract. Following a military coup known as the May Revolution in 1969 however, Sudan's government aligned itself with the USSR – and that was the end of the scheme. The situation of administrative non-interference changed only in the 1990s, when construction work began on the asphalted highway known as *Shiriyân ash-shimâl*.

The *Shiriyân ash-shimâl* Highway

Depending upon what is meant by the completion of a road – and meanings clearly differ – it is possible to state that the early 2000s saw the completion of the *Shiriyân ash-shimâl* highway. Based on a design survey from 1988 (which largely followed earlier stretches of road along the Wadi Muqaddam), earthworks began in the early 1990s with the construction of the roadbed. Without having officially been turned over to public use, as soon as the construction of the road reached the gravel layer – it was immediately used on a large scale by motor vehicles as a convenient gravel road, *radmiyya*. Similarly, during all the later stages of construction, parts of the highway were appropriated by drivers as soon as they were considered motorable, regardless of the damage their vehicles caused to the unfinished structure. Rehabilitation of the highway including the final sealing with a finishing asphalt layer was carried out between 2005 and 2007.

To understand its significance and meaning in the wider Sudanese context, the *Shiriyân ash-shimâl* highway should be considered as part of two large technological visions of the Sudan's future. The pressure from constituencies in Mahas and Dongola on central politicians to connect their areas to the capital has been high for quite some time, and there is a clearly discernible strategy of 'roads for votes' in the ruling party. Yet the highway would hardly have been realised if not tied to these visions. The first vision concerns the modernisation of the northern Sudan through a large-scale scheme of dams to produce hydroelectric energy for the electrification of the country and to irrigate vast areas for agricultural production. One of these dams, the Hamdab dam upstream of Merowe was completed in 2008, another one at Kajbar north of Dongola is under construction. Others at Dal, Abu Hamed, Shiraik and further sites are planned or under consideration (Verhoeven 2011). Large-scale displacements of local populations, in the range of half a million people, will be or have already been effected (Hänsch 2012). With regards to the highways, the *Shiriyân ash-shimâl* highway and another highway from Atbara to Merowe (completed in 2006), are used as feeder roads. The *Shiriyân ash-shimâl* was badly affected by the transport of heavy material to the construction site, and its rehabilitation in 2005 to 2007 was financed from funds allocated to the Hamdab dam project. It is evident that highway construction is an integral part of this developmental vision. According to a claim made by the Dams Implementation Unit, 1200 km of asphalted road (amounting to one quarter of Sudan's total length of asphalted roads) and four Nile bridges have been built to connect the region of the dams.[5]

The second vision ties into an even larger plan on a continental scale, namely the Trans-African Highway Project. The Trans-African Highway network is planned to reorder the interior African transport structure by connecting each capital of every African country through a highway network of close to 60,000 km, and thereby overcoming the legacy of the colonial transport structure which was largely based on extraction roads from ports to the hinterland (SWECO 2003: 7). These plans and their related funds adhere easily to the highly successful Sudanese policy aimed at accelerating the expansion of the

5 Sadd Marawî, the public relations magazine edited by the Dams Implementation Unit featured the roads and bridges constructed in connection with the Hamdab dam under the heading 'development conquers the heart of the desert' in its special April 2010 edition for the occasion of an inauguration ceremony at the Hamdab dam near Merowe (Dams Implementation Unit 2010: 72–77; 124).

highway network from the start of the century, which also coincided with the beginning of the Sudanese petrol boom. *Shiriyân ash-shimâl* is thus part of the Trans-African Highway's corridor 4 from Cairo to Gaborone and Cape Town. In the Sudan it meets Trans-African Highway's corridor 6 from Djibouti to Ndjamena and further west (SWECO 2003: 88). Although the Trans-African Highway is a planning and funding reality, *Shiriyân ash-shimâl* is not associated with any pan-African ideal by the Sudanese public, but like other road projects[6] it is known and imagined in compellingly national terms. The highly loaded concept of 'Artery of the North' implies that it establishes a vital connection in the national body between the centre at its heart and the northern Nile Valley.

The construction of the 'Artery of the North' is a state building activity in the sense of connecting the northern Nile Valley with the centre more generally, and in particular bringing the region of hydropower production and large-scale development of irrigation agriculture closer to the central region. After finishing the construction of the missing road links in northern Nubia (already close to completion in 2014), the 'Artery of the North' will also foster cross-border trade with Egypt by providing a direct connection other than the road through Halaib on the Red Sea coast and the deteriorating railway and ferry links from Khartoum to Wadi Halfa via Abu Hamed and across Lake Nubia (Lake Aswan).

Sudanese efforts of road construction from the turn of the century have been immense. Besides asphalted urban roads, the Sudan had one asphalted long-distance road of 200 km in 1980. Following the completion of the Khartoum to Port Sudan highway in 1982, the total length of asphalted long-distance roads amounted to 2,200 km compared to 16,000 km of gravelled road (Woodward 1991: 144). Since 2000 there has been a veritable explosion of highway construction across northern Sudan.[7] Sudan's petrol boom of 1999 provided the finance for road construction and building of other infrastructure (James 2011). On the other hand, the independence of oil-rich South Sudan in 2011, taking away one

6 Similarly, the designation *Al-Inghâz al-Gharb* – Salvation of the West – with its programmatic implications of national unity under the present *Inghâz* government, is used for the highway connecting Darfur.

7 The following roads have been constructed since 2000: Gedaref to Matamma, linking Sudan to Ethiopia, Port Sudan to the Egyptian border, the completion of the Berber to Abu Hamed road, the extension of the Kosti to El Obeid road to En Nahud, El Obeid to Kadugli, Atbara to Merowe, Omdurman to al-Multaqqa, al-Multaqqa to Dongola, Katranj to El Jinaid, al-Jabalain to Renk, Atbara to Haya and rehabilitation of Haya to Port Sudan, Merowe to Dongola via Karima and as-Silaim, Dongola to Wadi Halfa through Mahas, including an impressive number of bridges across the Niles. *Al-Inghâz al-Gharb* highway from El Obeid via En Nahud to El Fasher was close to completion by at the time of writing. And the El Obeid to Omdurman road had already reached Bara in 2015.

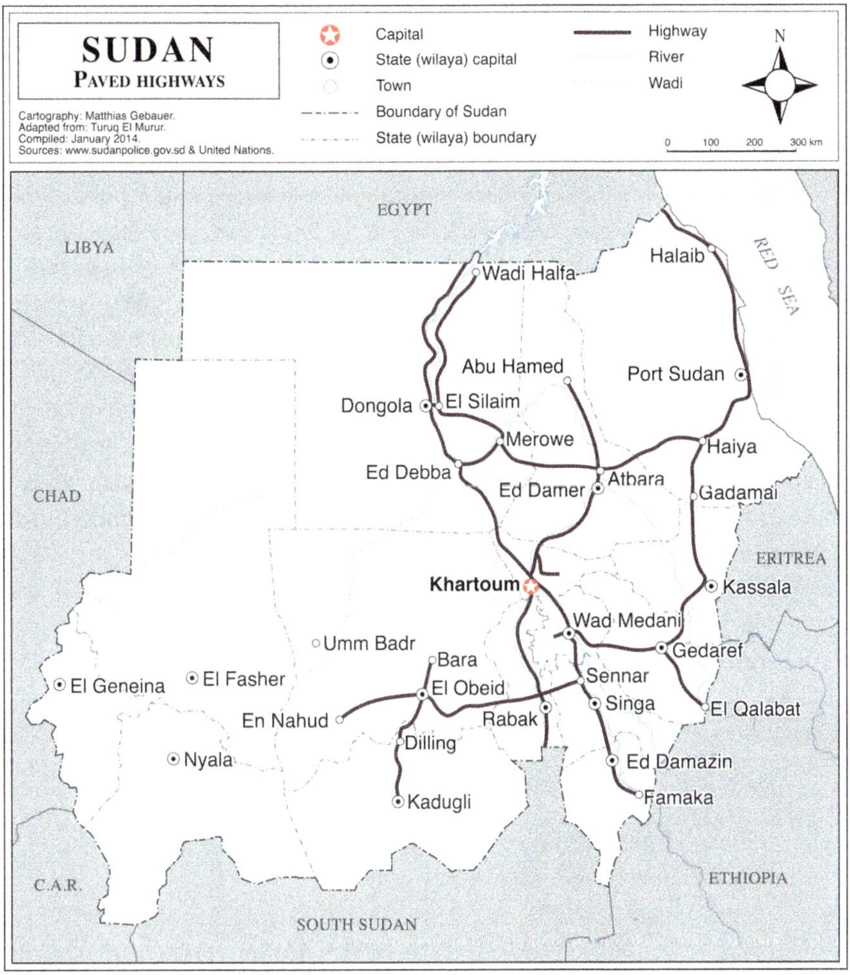

MAP 10.1 *Sudanese highways (2013)*

third of the territory and three quarters of the productive oil fields (Sidahmend 2013) as well as the ensuing fiscal retrenchment in Sudan from 2012 onwards will certainly slow down the expansion of the highway network.

According to a World Bank source, the highway network has nearly doubled from 3,400 km in 2004 to 6,200 km in 2008 (Ranganathan and Briceno-Garmendia 2011: 14). The accompanying growth of the trucking fleet in the early and mid-2000s amounted to 10% for lorries annually, 20% for tanker lorries and over 40% for dry cargo heavy trucks (Ranganathan and Briceno-Garmenida 2011: 11). The high growth rates for heavy trucks in particular are directly related to the expansion of asphalted overland roads, since articulated trucks do not travel earth tracks. Sudan's overall vehicle fleet increased from 485,813 in the year 2000 to 1,200,000 in 2007. During the same period, the number of road fatalities rose disproportionately fast from 492 to 2,200 (Ali 2010; WHO 2009: 195). This substantiates the impression that paving the roads invites death as an intimate travel companion. Paved highways regularly feature in the Sudanese press as virtual killing fields and have become a major public concern and this in turn provides a background for Sudanese concerns to implement international standardised norms of road safety (Ali 2010; Hissain 2013).

The overall regime of mobility in the Sudan is changing rapidly. By 2010, classified roads in the Sudan amounted to a total length of 31,300 km, of which 7,000 km were asphalted, 4,300 gravelled and 20,000 gravelled and earth roads.[8] Figures for non-classified roads, all of them earth tracks of the old Wadi Muqaddam type, are not available. It becomes evident from these figures that asphalted roads in the Sudan represent only a fraction of all roads. Automobility is thus situated in two distinct but interlocking technological regimes. In the older regime of unsurfaced and largely non-administered roads of the Wadi Muqaddam road type, and in the expanding technological regime of the planned and supervised tarmacked highway of the *Shiriyân ash-shimâl* type. Both regimes display their own mix of vehicles and their own travel practices. They are, however, interlinked with each other, and this provides space for tension and technological dramas.

The Regime of the Wadi Muqaddam Road Compared to the *Shiriyân ash-shimâl* Highway

When comparing the Wadi Muqaddam road with the *Shiriyân ash-shimâl*, the first difference to bear in mind is that the Wadi Muqaddam road has come into existence under the wheel. It has been made by its users, whereas the highway underwent a long gestation process of planning and rallying political support by weaving it into visions of a future Sudan, earmarking funds and finally, its actual construction. Users had no apparent part in the procedure. The entire process might have appeared as unpredictable as some of the stages of the emergence of the Wadi Muqaddam road did, but it was clearly dominated by coalitions of politicians, officials and civil engineers who tend to share a similar view as to what a modern road means and what its construction entails.

8 http://dlca.logcluster.org/display/public/DLCA/2.3+Sudan+Road+Assessment (accessed December 18, 2013).

The use of the road has changed fundamentally with the opening of the highway. For long-distance travellers using the *Shiriyân ash-shimâl* in transit, it has been transformed from a regime of desert travel on unimproved roads into a controlled regime of asphalted highways allowing for a smooth flow. Beside acceleration and concentration of traffic, this has entailed a different mix of vehicles and a functional differentiation of traffic, as well as an increase in the volume of traffic and in degrees of control.

Since the *Shiriyân ash-shimâl* highway became practicable for ordinary vehicles as a gravel road in the 1990s and as an asphalted road by 2000, the volume of traffic has increased considerably. In the 1980s the unimproved desert road through the Wadi Muqaddam was travelled almost exclusively by Nissan UD and Bedford TJ lorries and later Hino KY or XY lorries as well as by some 4X4 cars and buses manufactured in Sudanese workshops on the basis of a lorry chassis. With the construction of the highway, a fleet of articulated container trucks, mini buses and overland coaches entered the road. Most of the latter are Korean manufactured second-hand coaches owned by Dongolans, and since the extension of the highway towards the north, increasingly by people from Mahas. Many of these, through their body inscriptions, bear visible witness to the fact that their owners have earned the seed money for their businesses as migrants in the Gulf countries. Still, the volume of traffic on the highway is below 200 vehicles per day, though numbers rise considerably before and after public holidays. Heavy vehicles account for approximately 80% of total traffic.

This new mix of vehicles on the highway indicates an essential difference between transport on the unimproved road and on the highway. A clear functional differentiation of transport has taken place. On the unimproved road, lorries used to transport goods and passengers indiscriminately, loading passengers on top of the goods. In the highway regime, passengers travel by bus, sometimes in mini buses or saloon cars and goods are transported on trucks. This is clearly a matter of comfort, but it is also in line with the traffic laws of the Sudan which prohibit travelling on top of a loaded lorry, and which have recently been interpreted more strictly by the traffic police as part of the overall regulation of the highway regime.

In the regime of the unimproved road, there was not one single road, strictly speaking, but rather bundles of tracks that occasionally diverged far from each other. Drivers continually had to make decisions about which track to take as they picked their way around natural obstacles. Today, traffic is concentrated on the narrow band of black asphalt leading through the desert. The highway consists of a straight lane of smooth asphalt, with a carriageway of two times 3.5 m divided by lane markings and with gravelled shoulders of 1.5 m on either side. This allows for two trucks to pass each other at full speed, even next to a vehicle that has pulled over on one side of the road. Instead of constantly shifting through the low gears and forcing their rumbling lorries along the winding old tracks, drivers may now relax, their remaining excitement seems to derive from speeding and overtaking – which also keeps them from falling asleep, overworked as they frequently are. Skills tied to the practices of desert travel on the old roads such as the reading of the ground are now becoming obsolete with the new materiality of the highway, whilst new competences take their place.

In addition to lane markings, the new highway is also equipped with signposts and destination signs, the newly arising roadside settlements are signalled by signboards, such as, for example, Abu Dulû' (km 61), Bûhât (km 105), al-Ji'airîn (km 125), al-Qâ'aunâb (km 160), Tamtâm (km 190), Umm al-Hassan (km 238) and al-Multaqqa (km 303). Although nothing of the kind existed on the old Wadi Muqaddam road, unlike some rather notorious desert roads in other regions of the Sudan, getting lost was never a real problem except at night and during sandstorms. The *wâdî* was always nearby and its vegetation remained in sight of the driver along far stretches. Experienced drivers recognised landmarks that were also familiar to competent travellers and widely referred to in their conversations. Such landmarks included, for example, Jabal al-Haush, with its conspicuous archaeological site; challenging crossings of *wâdîs* running into the main *wâdî*; conspicuous clumps of vegetation, difficult stretches of sand dunes (such as Qauz Abu Dulû'), and clay pans such as Umm Jawâsîr, or nomads' wells. Drivers who have travelled the road extensively insist, however, that competent drivers simply knew their orientation from experience and that they required landmarks only for validation of their course or by way of emotional reassurance when in doubt. On the other hand, they report instances of complete disorientation, *râsî lâff* (Ar. spinning head), which they describe as a kind of disintegrating reality accompanied by disturbing feelings of such alienation that they would not even recognise familiar landmarks. Inexperienced drivers received simple advice, such as never to listen to the opinion of nomads when they referred to distances; always to be aware of which side of the *wâdî* they were; always to follow *al-lastik al-jadîd* (the fresh track); to lie down and sleep in case of *râsî lâff*; and if in doubt, to use a technique, which might be called reverse navigation. This is to follow the tracks of a vehicle coming from the direction in which oneself is heading on the assumption that the track has been proved to be manageable and will not end nowhere (compare Wadelnour, this volume).

To be a driver was a highly skilled and equally highly valued profession, as it came with the mastery of technological modernity and it required a long apprenticeship period. Under conditions of the unpaved road in rural Sudan, it is still regarded as such. Under the newly developing regime of the highway, driving has become a widespread middle-class skill. Individual drivers may rate their own driving skills very highly (and dispute those of others), yet it is not the mastery of skills that gives reason for admiration, but the possession of the car as an indicator of class.

The highway accelerates transport immensely. Its very materiality opens up potentialities for speed and flow that were unimagined under the old regime. On the Wadi Muqaddam road, it took a whole night and well into the next day for lorries from Omdurman to reach the bend of the river, and more often several days during the rainy season or when getting stuck in the sands of the notorious Qauz Abû Dulû'. The asphalted highway permits travelling at high speed. Overland coaches travelling at 100 km/h can now cover the 300 km in four hours (depending on time spent at tollgates, at roadblocks or for breaks) and small cars even less time than that. The smooth uninterrupted flow through the artery becomes the master expectation when entering the highway.

The Experience of Travelling the Desert Road versus the Highway

As travelling the unpaved road and travelling the highway are situated within different road regimes, the two forms are organised as different orders of the social. They rely upon different bundles of travel practices evoking different experiences and providing for different meanings. As such, the materiality of infrastructure becomes embodied through everyday practice. Strange as it seems in a place like the Wadi Muqaddam, the highway regime has turned the road into a space which French anthropologist Marc Augé (1992) has referred to as a 'non-place', a no man's land without history or meaning for the traveller; a senseless space, the only meaning of which lies in the fact that it has to be crossed.

Travellers in the modern overland coaches on the *Shiriyân* highway can be heard to remark in conversation: 'By God, the Sudan is really a vast country and look here at this wilderness! Can anybody live here?' Whilst in fact, the *wâdî*, some hundred meters away, is teeming with animals and people. Contemporary travellers take no notice because they are transported like in a cocoon. Functional differentiation, acceleration and the materiality of the highway combine to produce an experience of ecological isolation in traveller's minds, which has been unknown and even impossible when riding on top of a lorry. Travelling on a lorry led to a closer sensory interaction with the environment, passengers were exposed to local conditions, such as the heat of the day and the severe cold of the winter night, the dust, the wind and the movement of the lorry on the bumpy road, all of which left their marks on bodies and clothes (Wadelnour this volume). The passengers could feel, smell and see the landscape, with its climate, vegetation, soil, animals and people. During rests at the coffee houses operated by local nomads near their wells, they could establish contact and gain knowledge of the world in which they found themselves and they could learn the place names through conversation and know the desert from experience. Like the competences of drivers described above, these competences of passengers were also tied to practices of desert travel, and are becoming obsolete under the highway regime.

These days, passengers are instead transported in the same way as in an aircraft. Travellers complete their passage without ever being compelled to leave an urban Sudanese environment. The overland coaches are equipped with air conditioning, the window curtains are tightly drawn against the glaring sun outside, the seats are comfortable, and the chassis suspension absorbs every slight bump in the smooth asphalt cover. Bus attendants serve snacks and soft drinks, passengers doze off or watch videos on the monitor at the front of the coach. Passengers' sensory experience of and their engagement with the outside world is comparable to that of airline passengers – next to nothing. Only when the bus comes to a stop at a roadblock and the security officers burst in, do passengers become aware of an outside world again. They then try to realign themselves by pulling back the window curtains and, in a characteristic gesture, feeling for their mobile phones. This is the opportunity for hawkers outside to offer their goods through the windows. A second chance for contact with the outside world usually arises after two hours from Omdurman when the driver stops for a restroom or praying break and possibly for some tea or coffee either at Umm al-Hassan or the rest house of the *Shiriyân* company at Tamtâm. Other rest stops along the highway are almost exclusively visited by truck drivers and by local travellers.

This travel cocoon only cracks in cases of accident or serious breakdown. On the old road such incidences entailed long days of repair and waiting for spare parts fetched from afar. Having mechanical knowledge and skills was integral to being a competent driver and drivers and their assistants acquired such knowledge through apprenticeships and by participating in workshop maintenance and repair of their vehicles (Beck 2009; Hänsch 2009). Under the regime of the highway, the manager of the company will send a second bus in order to rescue his customers two to three hours after having been informed by mobile phone.

Passengers disembark at their destination as clean and relaxed as they entered the bus in the morning. When they used to travel on lorries, they prepared themselves carefully for the journey by bringing blankets and provisions. The lorry driver made sure to carry water as well as an assortment of flour, spices and cooking equipment. He stopped at the coffee houses in the *wâdî* for tea, and passengers shared their meals with each other. Most of the old coffee houses used to sell only tea and coffee, not meals or milk, as the nomads regarded the selling of food as *harrâm*. Instead, if travellers met a herd of animals at a well, they could claim milk as a gift and the driver would buy chicken or a lamb, which his assistants with the help of some passengers prepared during a rest. In the rest spots along the highway, the selling of food is not perceived as a problem. In fact, commoditisation spreads along the highway. After the precedence set by the rest houses of the *Shiriyân* company, restaurants and coffee houses sell urban style meals of beans or meat to lorry drivers and to other travellers who bother to stop. Nomads sell animals to be prepared in the roadside restaurants in exchange for imported commodities, which is part of the interlocking economies along the highway. As far as I am aware, there is only one coffee house along the highway, where the old practice still persists. It belongs to a very old lady near the small settlement of Umm al-Hassan who learned the trade as a small girl from one of the most famous coffee house-keepers in the area, Umm al-Hassan, after whom the settlement took its name. Only regular customers pull up at her old-fashioned grass hut, in particular lorry drivers who make it a point to stop there for a calm nap whilst fleeing the noisiness of other rest stops.

The earlier practice of eating and preparing meals in groups was called *halla*, after the common Sudanese designation for a cooking pot, but the word extended beyond commensality to include companionship, for instance to mean the particular form of sociality of migrant labourers embarking together on their journey and staying together as a group. Sharing meals, the organisation of buying and preparing food, the shared effort of digging out a lorry bogged down in the sand, mutual assistance of all kinds, the organisation of space on top of the load as well as the awareness of shared hardship and common destination tended to transform random individual passengers into a community of companions with a particular social order.

In the comfortable buses, distantly related forms of sociality may emerge, although as a rule these are much more casual, resembling the kind of travel acquaintances in Georg Simmel's (1903) fleeting sense rather than travel companions in the sense of *halla*. Bus passengers do not contribute to the making of the journey in the same way that the crowds on the lorry did. They are transported without having a hand in the journey or an active commitment to its success. Bus passengers may chat with their neighbours ('can anybody live in this wilderness?'); hand over their newspaper or occasionally offer sweets, but there are neither common tasks nor shared responsibilities. And beyond small civic gestures of friendly attentiveness, such as passing over somebody's baggage or ceding space for a neighbour's elbow, interaction is of a non-committal nature. The social order of the bus is radically different from the social order of the lorry. Whereas travel companionship is grounded in a solid assumption of shared effort and mutual support, travel acquaintances are free from obligations.

Similarly, the highway regime reduces the outside space to meaninglessness for the transient traveller. Whilst reading his newspaper, watching the movie on the front screen or simply dosing off, for the bus passenger in his cocoon the Wadi Muqaddam ceases to exist as a social space. It could be anywhere or, for that matter, in outer space. For transient travellers it evokes no emotion and no experience, except boredom.

Roadside Communities and the Reordering of Space

The Wadi Muqaddam and its adjacent mountain ranges are traditionally inhabited by nomads, who claim the clay pans, wells and pastures as their tribal homelands (Ar. *dâr*). Because of its reliable underground water, the Wadi Muqaddam area has been known since early times as 'the desert of Gabra with its 100 wells' (Budge 1907: 392), named after the main wells at Jabra Sa'îd located on the border of today's Khartoum and Northern States, most of which are in the *wâdî* itself or its tributaries. The area northeast of Omdurman is inhabited by Kabâbîsh, belonging to the Barâra branch. West of them, near the wells of Abû Hât, and already across the borders to Kordofan, live Hawâwîr, mainly of the Jautâb branch. To their north live Qiraiyât around their wells of Jabra Sa'îd, Umm Haraut, Bîr al-Akhann, Fanqa and Jî'airîn as well as Hassâniyya at Bîr al-Hassâniyya and al-Biraira. North of Bîr al-'Ajamî, at the intersection of North Kordofan, Khartoum and Northern states, live Hawâwîr, mainly of the Rûbâb branch, at their main wells of Bîr Aminalla, Umm Rimaila, Umm Rataut and al-Ikhrît near the recently built town of Tamtâm, and from there north at Bîr al-Haush, Umm al-Hassan and Umm Taub all the way to the Nile, with a small enclave of Sawwârâb near the wells of Umm Jaqqarî and Adât Matar. This nomadic space tends to become reordered through the highway.

The highway attracts populations and resources which had previously been dispersed throughout the *wâdî* and its hinterland. The construction of the highway included the building of mosques and dispensaries as well as schools at places where roadside settlements have emerged. It also included the construction of two model towns (Tamtâm and Umm al-Hassan) that were planned with the dual purposes of settlements for the Hawâwîr nomads and as service stations for the road traffic. Petrol stations and rest houses for travellers, owned and subcontracted by the *Shiriyân ash-shimâl* company, were built at al-Multaqqa near the river and Tamtâm halfway from Omdurman. Umm al-Hassan and Tamtâm have now become administrative and trading as well as service centres. Those Hawâwîr, who had laid the foundations of a mud house and planted trees within their enclosure, received 30,000 Sudanese dinars (at that time roughly US$ 115) as an incentive to construct a house and settle in Tamtâm.[9] In 2013, Tamtâm had approximately 100 inhabitants, Umm al-Hassan half this number. All of them made their living from the road, whether directly servicing the road as traders and hawkers, operators of restaurants and coffee houses, tyre repair and mechanical workshops, or indirectly as teachers, medical dressers, watchmen or labourers working in the irrigation of the attached afforestation project at Tamtâm. Transmission towers for the cellular phone network of three Sudanese network providers run alongside the highway at distances of 30 km between each other. Each of the thirty towers offers an income for the families of their guards. The nomads tend to understand the mosques, dispensaries and schools – which were erected in connection with the highway – in terms of compensation by the state and company in return for their land that they have lost to the road.

Those 1,500 to 2,000 Hawâwîr who belong to the administrative areas of Tamtâm and Umm al-Hassan and who do not reside in the towns prefer to live at a distance of several kilometres from the road, such as in the areas of Umm Jawâsîr or al-Haush on the fringes of the Wadi Muqaddam. If they need to travel or if they require goods like fresh bread or credit for their mobile phones from al-Multaqqa, they call on a daily minibus that operates between Tamtâm and al-Multaqqa, whose very enterprising young driver picks up his passengers or drops their desired goods at prearranged sites near the road. Nomadic families may send their herds to pastures at a distance of up to thirty to eighty km from the road. It has always been common practice among the Hawâwîr

9 The Norwegian anthropologist Kjersti Larsen provides an account of how a development project nearby, at Umm Jawâsîr lost some of its appeal as a place of settlement for the Hawâwîr because of the growth of Tamtâm (2008: 80–81). Umm Jawâsîr had been implemented as an irrigated resettlement project for those Hawâwîr who had migrated to the Nile valley after having lost their herds in the severe drought of 1983–85 (Larsen 2002, 2003, 2007; Haug 2000). After Norwegian development funds had been phased out, Umm Jawâsîr was completely devastated in 2013, earth parched and palm trees dead. Not one of the water pumps was functioning, and the whole project deserted. I suspect that it was less Tamtâm but the phasing out of funds which was responsible for the desertion of Umm Jawâsîr.

to live very dispersed in small camps of two or three closely related families near one of their wells and to release their camels for pasturing, unattended by herdsmen and only keeping mares in milk and small livestock near the camps. When they are thirsty, the camels will come back to the well where the herdsman can receive and water them. Only large herds are usually accompanied by herdsmen, and these tend to stay beyond the western mountain ranges in Northern Kordofan. This arrangement of unattended pasturing will only work if stray camels are restored back to their owners and thieves prevented from stealing them, which implies that the whole area should be under corporate control, and this is one of the reasons why the Hawâwîr tend to guard their tribal homeland as jealously as they do (Beck 1988: 229).

Even if nomads live two day's walk away, they are drawn into the orbit of the highway, as they come down to the roadside shops from time to time. They may bring a lamb or some goats which they trade for tea, sugar, flour and possibly some medicine. After having exchanged the latest gossip over tea and spent the night on the shop's veranda, they disappear again in the desert with their provisions tied to the saddle of their camels. The 'Artery of the North' thus serves as the main artery of the region's economy with a capillary network branching off into a large but sparsely inhabited nomadic hinterland.

The tribal territorial regime and the regime of the highway are clearly interconnected. Sometimes, tension arises as the new highway opens up opportunities for the reordering of the space, for instance, for the economic transformation of agricultural land. Large tracts of what looks like desert to outsiders, but is in fact pastureland, tends to acquire a new value for investment, as it comes within easy reach of the capital. Large-scale poultry farms, feedlots and dairy farms; irrigated vegetable and fruit plantations for providing food for the capital's inhabitants already line the highway up to fifty km from Omdurman. According to the authority of the Sudanese Land Act, tribal *dâr*-rights, being of a non-registered legal kind are not likely to be upheld by courts, and, backed by high party and government officials, commercial interests have succeeded in appropriating and fencing in large tracts of land in the Qiraiyât and Barâra areas along the highway. Sometimes, however, this form of land grabbing has also been successfully resisted by tribal communities in acts of self-defence which are only imaginable in fragmented states like the Sudan, where parallel and shadow structures contest each other's and the formal structure's authority. A locally famous case of resistance involved a community of Hawâwîr that was backed by a tribal force against the investor, the then-Governor of Khartoum State who was backed by a police force. The situation threatened to result in an armed showdown and necessitated the first Vice President to be flown in by helicopter to defuse the conflict and to negotiate a retreat. The incident was obviously intended as a test case by the Hawâwîr leadership and is therefore not likely to be repeated on a regular basis, but it highlights the extent of roadside friction produced at the intersection of the highway and the tribal regime. The highway has thus provided the potential for a reordering of the political economy of the whole area – but that is a political process, which means that there is no techno-teleological automatism involved. Given the strength of the tribal territorial regime, and of the weakness of the state's institutions, the construction of the highway does not automatically lead to primitive accumulation of this kind, yet it generates distinctive affordances for it.

The Tentacles of Government

The highway is part of a larger technological regime, as already remarked, it is more than simply a strip of asphalt. Its materiality provides the potential for the reordering of the political economy and it has its place in regional politics and in the larger visions of the modernisation of the Sudan, as explained above. Furthermore it is also a 'path of authority' (Fairhead 1992), the tentacles of government spread along it. Being a toll-road operated by the private company of the same name, access to the *Shiriyân ash-shimâl* is regulated by tollgates where the highway enters the area of sand dunes at the settlement of Qauz Abû Dulû' and near al-Multaqqa at the other end. As such it comes under the Law for the Regulation of Tollroads (Government of the Sudan 1980). The National Security and Intelligence Service maintains a notorious roadblock (*taftîsh*) at al-Multaqqa where vehicles can be thoroughly searched, and officers and informants are dispersed in the roadside settlements. An army camp for the defence of the capital is situated directly at the roadside on Qauz Abû Dulû' and a second one at the trench which surrounds the capital,[10] where there is also another *taftîsh*. Al-Multaqqa is heavily guarded as well. Being part of the Trans-African Highway Network and of Sudan's highway system, the *Shiriyân ash-shimâl* comes under the Sudan Road Traffic Act of 2010 (Government of the Sudan 2010) as well as the Law on the Protection of National Roads of 1994 (Government of the Sudan 1994) and is under control of the National Highway Authority and the National Highway Police. The highway

10 The highway is guarded more heavily after a surprise attack by rebels of the Darfur based Justice and Equality Movement, who managed to drive 130 armed vehicles through the desert from Darfur and on 10 May 2008 forced their way into Omdurman on the highway. 'Path of authority', yes, but also path of rebellion.

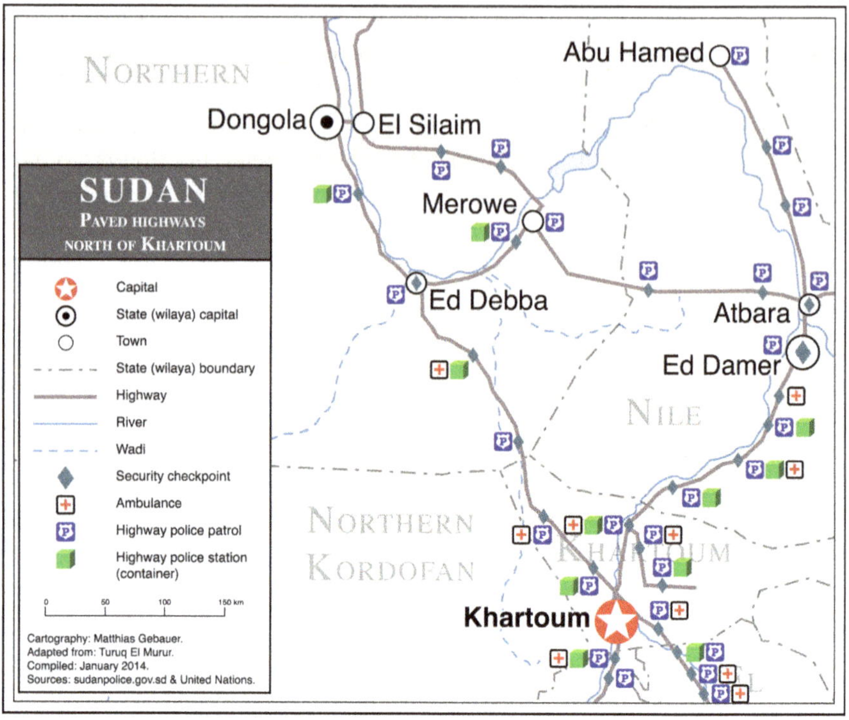

MAP 10.2 *The regulatory infrastructure of the highway*

police maintains stations in Omdurman and near al-Multaqqa. Highway police vehicles are stationed at the toll gate in Abû Dulû', at al-Qâ'aunâb and at Tamtâm where the highway police also runs an office. Ambulances are stationed at al-Multaqqa and Abû Dulû'.

Planners and state institutions make concerted efforts to include the paved highway in a particular technological zone that is highly controlled in order to impose specific technological norms. This is partly in response to advice and pressure from international agencies – for instance in connection with the Trans-African Highway Program – who work to expand the zone of international technological standards[11] and partly in response to a large public concern with road safety in the Sudan itself. Besides structural designs and construction materials, technological standards include norms regarding,

11 As materialised in the Inter-governmental Agreement to underpin the Trans-African Highway Network of 2013 after the precedence set by the European Agreement on Main International Traffic Arteries of 2001 and the Intergovernmental Agreement on the Asian Highway Network of 2005 (Mutangadura 2013). See also the publications and recommendations on worldbank.org/afr/ssatp/Publications.

among many others, the maximum axle load of trucks, roadworthiness, third party liability, general driving behaviour, maximum speed, safety belts, child retention devices, alcohol and narcotics (ab)use and pre-hospital assistance following accidents. Practices of control are elaborate and materialise in passenger lists, license plates, ID cards, insurance tags, registration numbers, driving licenses, tickets issued at tollgates and so forth. Some of these are routinely inspected while others are shown on demand. The vision is a finely supervised modern technological regime, originally crafted in Europe and then extended to the level of international standards, which is compatible and connectable across international borders, and which promises a smooth flow.

Under the regime of the unimproved road on the other hand, anybody who has a vehicle is free to participate, because the highway's technological norms are either not implemented or not enforceable. Being free to participate does not mean that any vehicle is able to successfully participate, of course. There are valid reasons for the elaborate technical adjustment of desert-going vehicles (Beck 2009) and there is a code of conduct on the unimproved road (Wadelnour this volume). Nonetheless, overloading lorries, non-compliance with standards of roadworthiness, carrying passengers on top of the freight and similar practices are considered normal and are taken for granted. On the old unpaved road outside of the towns, nobody used to care – or put more correctly, there used to be no authority to care. Yet, now entering the highway regime from the regime of the unimproved road can resemble entering a foreign country without a passport. This, at least, is how local people have described the experience. Being rooted in the regime of the unsurfaced road, both drivers and their vehicles become the undesirables of the highway regime. Their vehicles, in which so much effort has been invested to adapt them to the conditions of the unimproved road (to make them 'desertworthy', if that neologism is allowed) no longer meet the standards of the highway regime. It is the very modifications that make vehicles desert-going and adjusted to the regime of the unimproved road – chassis extension, adding an additional axle, strengthening of wheel rims, replacement of nuts and bolts, wheelbase alterations, strengthening of the chassis (thereby occasionally tampering with chassis numbers), additional spring leaves, conversion of automatic to manual transmissions, even conversion to another type of vehicle, all of them without certification by the original manufacturer or technical vehicle inspection – that also make vehicles illegal on the highway. Visible appearances often betray the roadworthiness of ancient Bedford lorries and of equally aged Toyota pickups. These vehicles have no seat belts; the steering of lorries may allow for too much play (which is well suited for dirt tracks but creates problems with directional stability on the highway); the tread depth of their tyres is considered less than enough; their overload is excessive; they carry passengers by the dozen, and child retention devices are totally unheard of. Registration tags, which have to be renewed annually according to the Traffic Act of 2010, have expired for quite some time and driver's licences – if any – are worn out by rough handling and have become undecipherable over time. Particularly infuriating are the experiences of the tribal owners of the land, when they are required to pay an entry fee into their own country at the tollgate, which they tend to regard as a totally unreasonable demand. The highway embankment then acquires all characteristics of a border, and is, indeed, the invisible border between two regimes of automobility.

Technological Dramas, or Who Makes the Order of the Road?

The border provides a scene for all sorts of technological dramas regarding access. In reality, these are not simply technological dramas, they only appear as such if their perspective has been accepted as a truth. As in other cases, technology here masks politics, and as Helmut Schelsky (1954: 25) claimed: 'Against the technologically established truth every opposition is irrational.' Regulating bodies design the norms of the highway regime in a way that naturalises ('technologises') access in a technological, non-political rationality serving the vision of a safe and uninterrupted smooth flow. What entails is that those vehicles that do not meet the norms are reconstituted as an obstacle and as an anomaly, even a pathology, of the road, which has to be eliminated. Technological dramas are therefore also political dramas. They are about participation and about exclusion of those who do not wish or who cannot afford to participate on these technological terms. The latter are compelled to find surreptitious (and often quite humiliating) ways if they still want to participate – yet, as becomes clear, they do find ways, through practices of evasion.

The magnitude of the border problem is plainly stated in a recent assessment report on the consequences of the Sudan Traffic Act of 2010 for road safety. According to the author, a high-ranking officer of the Sudanese Police Force, less than half of the estimated 1,200,000 motor vehicles in the Sudan underwent the legally required annual registration and technical examination between 2009 and 2011.[12] It is certainly no misinterpretation of the figures to

---

12 According to the database of the General Directorate for Traffic, only 610,307 of the estimated 1,200,000 vehicles (year 2008) were registered in the year 2009, followed by 442,495 in 2010 and by 461,006 in 2011. The author notes the weak control of the General Administration of Traffic on those that are actually in use on the road, and hence their weak coverage by the technical examination which is to ensure their roadworthiness (Hissain 2013).

suggest that the majority of these unregistered vehicles are put to use on unimproved roads in rural areas, where control is singularly weak compared to urban areas and under the highway regime. The regime of the unpaved road is the natural habitat for those vehicles that escape registration and technical examination.

Given the high figures for the vehicles escaping being controlled, it would be unrealistic to suggest that all the standards enumerated above are actually met on the highway by all vehicles despite administrative efforts to implement them. Transit traffic seems comparably easy to control as access to the highway leads through the tollgates at both ends. The sheer materiality of the new long-distance coaches, articulated trucks and saloons does not afford itself to circumvent control on non-paved detours. Short of erecting a fence along the road, however, local traffic seeping in along the 300 km long desert flank of the asphalted lane from the regime of the unimproved road cannot be easily fended off. Seeping in, in fact, represents a mass movement, the 'quiet encroachment of the ordinary' (Bayat 2010), unorganised but on a large scale. As might be expected, large spaces of discretion between the different road regimes have evolved, and navigating the fringes of the highway's technological zone involves much struggle and conflict, negotiation and complicity, evasion and bending of rules.

Schemes that are made up in intergovernmental committees to ensure internationally standardised norms are thus not able to stand their test on the asphalt. It is evident that they do not make the road's order. Yet, they have consequences once the machinery of control is set in motion. Security checkpoints, the possibilities of baggage control and personal searches, tollgates as well as supervision by traffic police combine to make everybody who wants to participate in the order of the highway regime feel under surveillance. As a consequence, the zone of discretion opens large possibilities for small cash to change many hands. For instance, well-meaning bus drivers are sometimes tempted to overload their buses by seating passengers (especially those whom they pick up on the open road) on the water barrel, baggage in the aisle or on the steps by the exit. This is not normally a problem with officers of the National Security and Intelligence Service, who just check the passengers' names against the passenger list and who are not particularly concerned with offences against the Traffic Act. The driver is, however, well advised to wrap a bank note as 'coffee money' into the passenger list when he sends the bus attendant to present it to the traffic police. As can be imagined, coffee money and similar practices are well established to ease negotiation in the zone of discretion which connects the two different regimes.

Detours become common practice as well. Local vehicles usually travel at a slow pace and are easily recognizable by their appearance, which bears witness to their biography of desert travel. *Mahallî*, 'local', a passenger might comment to his neighbour in a rare moment of attention to what happens outside, as their bus roars past the slow moving obstacle. As soon as a checkpoint, tollgate or highway police car stationed on the highway's shoulder comes into sight, transit vehicles slow down while the *mahallî* vehicle will simply leave the highway on much used tracks that conspicuously fan out into the desert. Once beyond the obstacle, it will return back onto the highway again. *Mahallî* thus emerges as a distinctly recognizable category of its own in the order of the highway.

It would be wrong, however, to interpret the detour and generally the zone of discretion as practices that have evolved entirely naturally. Like with other practices, it would be more appropriate to regard them as historical outcomes of social experience and in the case at hand, as outcomes and part of the technological dramas, which in the long run shape the social order embracing the road and the roadside.

Roadside commentators point to a long history of being coerced to pay their way, usually being rather vague as to whether the payment in question should be regarded as a bribe or as a fine. Most likely, this distinction is irrelevant for them anyhow, as the payments are widely understood as a kind of tax among local people. This matches the implicit and equally widespread perception of officers who reside at the roadside settlements, like a kind of occupation army, with whom there might be complicity but no familiarity. Officers belonging to the national police force are strangers to the local roadside community, and they remain strangers, even if they have spent long times of service in the so-called *manâtiq ash-shidda* (Ar. hardship areas), a kind of Sudanese Siberia. In spite of being powerful strangers, they feel particularly vulnerable to ostracism because of their isolated position in the roadside order. They may perceive transit traffic as easy targets but with local traffic, *mahallî*, they are well advised to apply some restraint and a good measure of road-level discretion.[13] Police officers (together with their families, if they do not live as bachelors,

13 The term is appropriated from Michael Lipsky's classic 'Street-Level Bureaucracy' (2010 [1980]) where the author analyses the dilemmas of individuals in public services and shows how envisaged policy actually comes to differ from policy enacted on the front line of bureaucracies through what he aptly designates as street-level discretion. These dilemmas were acknowledged by the author of the assessment report cited above in an interview (March 11, 2015). Yet, from his perspective as a Police Major General he explained traffic police practice as resulting from a mixture of good-natured Sudanese custom, limited technical means and extremely low salaries.

'*azzâba,* which makes their situation even more precarious) inevitably have to get along with the local community off-duty as well, which means that their goodwill in coming to terms with what they have been trained to regard as offences during hours on duty will be taken into account. In circumstances such as these, being characterised as *dukhrî,* straightforward, should not be taken as a positive social recommendation, and in a desert outpost like Tamtâm or Umm al-Hassan, prospects of being ostracised by the local community appear particularly threatening.

Technological dramas shape technology-as-lived-with as distinct from technology-as-devised. They are usually enacted upon the small, mundane stages of everyday life along the road. Although the particular dramas may be long forgotten, they have still produced their own significations such as *mahallî* and more or less taken for granted guides for acting. They are as much part of highway users' competences as reverse navigation is in the regime of unpaved roads (see above). The dramas and their outcomes establish bundles of practices, which in the end constitute the order of the road. To be fluent in this mundane order of the road is part of the craft of travelling as much as of the craft of policing the road.

If this order is seriously challenged however, whether by divergent interpretations or by attempts at unilateral transformation, or be it simply from incompetence, small technological dramas may culminate in violent crises. This is illustrated in the following story, which is well known along the road. As the story goes, one day a trader from the river passed the highway police station just outside of al-Multaqqa in his pick-up without stopping. Supposedly, he did not intend to travel the highway as he was heading to a place in the desert where he had some business. The police followed him in two cars and when he did not stop, they opened fire on the pick-up. Some bullets hit the cabin and one penetrated the door and the trader's leg. Word of the incident spread rapidly by mobile phone, and by the afternoon, a large crowd from the river and from all along the highway had gathered in front of the police station, with others still arriving. The crowd staged a large demonstration and even the reserve police called in as reinforcement did not dare confront the angry crowd. It was only after a high government official arrived at the scene as a mediator that the crowd dispersed the following day and the drama came to an end. 'Since that day', the narrator concluded, 'they do not ask us. It is as if they did not see us. Yet, you have to leave the road and drive around his back. Of course, you later return to the highway again – and still he will pretend that he does not see you. Yet you have to respect him. Never pass in front of him!'

Whatever the exact historical truth in the content of the story (and I have come across slightly diverging versions), its significance lies in its function as a corroborative myth for the practices of discretion. It can be told and retold and it lays an exceptionalist foundation for the *mahallî* status, making it a strong counterstatement to the master expectation of uninhibited safe flow, roadworthiness and the undesirables of the highway regime.

In conclusion, it would be a serious categorical misunderstanding to take the Traffic Act as plainly constituting the order of the road, even in the regime of the highway. Sociologically speaking, nothing could be further from the actual situation at road-level, even if that sort of wishful thinking seems widespread in intergovernmental committees and agencies and generally among adherents of social engineering. The attempts at a legal implementation of a particular technological zone and its enforcement have to withstand their test on the road itself, as becomes evident in more sober professional accounts (see Hissain 2013). Enforcement is neither without reservation, as it is conditional on the strength of other players, a strength that may materialise in all kinds of friction, deviation, the quiet encroachment of the ordinary, negotiation, complicity, or, indeed, outright violent confrontation. All of these are modes of the technological dramas that are enacted along the road. The *Shiriyân ash-shimâl* story may thus shed some light on the implementation of technological zones and generally on the transfer of technology from the North Atlantic to the global South through the level of internationally accepted standards.

Keeping in mind that all orders are temporary orders, the road-level order achieved through all these dramas might well turn out to be a temporal compromise, and as such, only a stage in the general transition to a modern highway regime as designed in intergovernmental committees and based upon international norms. On the other hand, the temporal practical compromise might also stabilise into a sustainable order. This is highly probable as the basic contradiction between the regime of the highway and the regime of the unimproved road, which is at its heart, is not likely to be resolved in the near future. Time will tell. What will happen depends on whether fiscal retrenchment after the booming years of the Sudanese economy (and state likewise) comes full circle, with the consequence that the road starts to deteriorate and the state has to cut salaries and resources in the highway police force. We can then even imagine a third outcome, namely the highway regime being swallowed by the regime of the unsurfaced road, as cracks and potholes start to dominate the highway and sand is eventually blown over the remnants of asphalt and deserted tollgates, making the road motorable again for desert-going vehicles only.

References

Ali, Galal. 2010. 'Traffic Accidents and Road Safety Management: A Comparative Analysis.' In *Proceedings of the 29th Southern African Transport Conference (SATC 2010)*, 530–540. Pretoria.

Augé, Marc. 1992. *Non-lieux: Introduction à une anthropologie de la surmodernité*. Paris: Éditions du Seuil.

Barry, Andrew. 2006. 'Technological Zones.' *European Journal of Social Theory* 9 (2): 239–253.

Bayat, Asef. 2010. *Life as Politics: How Ordinary People Change the Middle East*. Stanford: Stanford University Press.

Beck, Kurt. 1988. *Die Kawâhla von Kordofan*. Stuttgart: Steiner

Beck, Kurt. 2009. 'The Art of Truck Modding on the Nile (Sudan): An Attempt to Trace Creativity.' In *The Speed of Change: Motor Vehicles and People in Africa, 1890–2000*, edited by Jan-Bart Gewald, Sabine Luning and Klaas van Walraven, 151–173. Leiden: Brill.

Beck, Kurt. 2013. 'Roadside Comforts: Truck Stops on the Forty Days Road in Western Sudan.' *Africa* 83 (3): 426–445.

Betz, Mathew and Richard Baumann. 2007. 'Planning Roads for Darfur, Sudan.' *Transportation Research Record, Journal of the Transportation Research Board* 1989: 181–192.

Budge, E.A.W. 1907. *The Egyptian Sudan, its History and Monuments*. London: Kegan Paul.

Dams Implementation Unit. 2010. Accessed July 18, 2016. http://www.merowedam.gov.sd/mag/74/.

Fairhead, James. 1992. 'Paths of Authority: Roads, the State and the Market in Eastern Zaire.' *The European Journal of Development Research* 4 (2): 17–35.

Government of the Sudan. 1980. *Qânûn rasm istiʿmâl turuq al-murûr al-sarîʿ li-sannat 1980*. Accessed July 18, 2016. http://moj.gov.sd/content/lawsv4/3/20.htm.

Government of the Sudan. 1994. *Qânûn himâyat at-turuq al-qaumî li'l-sannat 1994*. Accessed July 18, 2016. http://www.moj.gov.sd/content/lawsv4/7/18.htm.

Government of the Sudan. 2010. *Qânûn al-murûr li'l-sanat 2010*. Accessed July 18, 2016. http://trafficpolice.gov.sd/index.php?option=com_content&view=article&id=140:2013-02-10-11-57-41&catid=49:2012-11-03-19-06-46&Itemid=11.

Grant, J.A. 1875/6. 'Itinerary from Debbé to El Obeyad, communicated by General Stone, Chief of the General Staff, Egyptian Army.' *Proceedings of the Royal Geographical Society of London* 20 (4): 357–362.

Hänsch, Valerie. 2009. *Sifinja: The Iron Bride*. Documentary DVD, Germany, 70 mins.

Hänsch, Valerie. 2012. 'Chronology of a Displacement: The Drowning of the Manasir People.' In *Nihna nâs al-bahar: We are the People of the River. Ethnographic Research at the Fourth Nile Cataract Region, Sudan*, edited by Cornelia Kleinitz and Claudia Näser, 179–228. Wiesbaden: Harrassowitz.

Haug, Ruth. 2000. *Livelihood Security among Pastoralists in Northern Sudan: Post-Hunger Development in a Country at War*. Noragric Working Paper 19. Aas: Centre for International Environment and Development Studies.

Hill, Richard. 1965. *Transport in the Sudan*. London: Oxford University Press.

Hissain, A. 2013. *At-tashrîʿat wa atharhâ fîʾl-hadd min al-hawâdith al-murûrîya. Dirâsa tahlîlîya li-qânûn al-murûr as-sûdânî li-sannat 2010*. Accessed September 17, 2015. http://trafficpolice.gov.sd/images/tshre3.pdf.

James, Laura. 2011. 'The Oil Boom and its Limitations in Sudan.' In *Sudan Looks East: China, India and the Politics of Asian Alternatives*, edited by Daniel Large and Luke Patey, 52–69. London: James Currey.

Khogali, Mustafa. 1973. 'Sudan Transport and the Development of Industry.' In *Industry in the Sudan*, edited by F. Imam. Khartoum: Government Printing Press.

Klaeger, Gabriel. 2012. 'Rush and Relax: The Rhythms and Speeds of Touting Perishable Products on a Ghanaian Roadside.' *Mobilities* 7 (4): 537–554.

Larsen, Kjersti. 2002. *Mobility, Identity and Perceptions of a Good Life: The case of the Hawawir, Northern Sudan.* Oslo: Skrifter Universitetet i Oslos kulturhistoriske museer.

Larsen, Kjersti. 2003. 'Mobility, Identity and Belonging: The Case of the Hawawir, Northern Sudan.' In *In the Maze of Displacement: Conflict, Migration and Change*, edited by N. Shanmugaratnam, Ragnhild Lund and Kristi Anne Stolen, 106–127. Kristiansand: Høyskoleforlaget.

Larsen, Kjersti. 2007. 'Adaptability, Identity and Conflict Mediation among the Hawawir in Northern Sudan.' In *Conflicts Over Land and Water in Africa*, edited by Bill Derman, Espen Sjaastad and Rie Odgaard, 75–93. London: James Currey.

Larsen, Kjersti. 2008. 'Far from the Battlefield: Livelihood Revival, Settlement and Modernization among the Hawawir, Northern Sudan.' In *Between War and Peace in Sudan and Sri Lanka*, edited by N. Shanmugaratnam, 77–93. London: James Currey.

Lebon, John. 1955. 'Desert motor routes in northern Sudan.' *Geography* 40 (4): 271–3.

Lipsky, Michael. 2010 (1980). *Street-Level Bureaucracy: Dilemmas of the Individual in Public Services.* New York: Russel Sage Foundation.

March, M. (1948). 'Transport in the Sudan.' In *Agriculture in the Sudan, Being a Handbook of Agriculture as Practiced in the Anglo-Egyptian Sudan*, edited by John Douglas Tothill, 176–186. London: Oxford University Press.

Martzsch, Ursula. 1939. *Moderne Verkehrswege im tropischen Negerafrika.* Würzburg: Triltsch.

Morrice, H.A. 1949. 'The Development of Sudan Communications, Part 2.' *Sudan Notes and Records* 30: 141–175.

Mutangadura, Gladys. 2013. *The Inter-Governmental Agreement to Underpin the Trans-African Highway Network.* Washington: World Bank.

Pfaffenberger, Bryan. 1992. 'Social Anthropology of Technology.' *Annual Review of Anthropology* 21: 491–516.

Ranganathan, Rupa and Cecilia Briceño-Garmendia. 2011. *Sudan's Infrastructure: A Continental Perspective*. World Bank Policy Research Paper 5815, Washington: The World Bank.

Schelsky, Helmut. 1954. 'Zukunftsaspekte der industriellen Gesellschaft.' *Merkur* 8: 13–28.

Scott, James. 1998. *Seeing Like a State: How Certain Schemes to Improve the Human Condition Have Failed*. New Haven and London: Yale University Press.

Sidahmed, Alsir. 2013. 'Oil and Politics in Sudan.' In *Sudan Divided: Continuing Conflict in a Contested State*, edited by Gunnar Sorbo and Abdel Ghaffer M. Ahmed, 103–120. New York: Plagrave Macmillan.

Simmel, Georg. 1903. Soziologie des Raumes. *Jahrbuch für Gesetzgebung, Verwaltung und Volkswirtschaft im Deutschen Reich* 27 (1): 27–71.

Sudan Government. 1904. *Reports on the Finances, Administration and Condition of the Sudan 1903*. Khartoum: Sudan Government.

Sudan Government. 1906. *Report on the Finances, Administration and Condition of the Sudan 1905*. Khartoum: Sudan Government.

Sudan Government. 1930. *Report on the Finances, Administration and Condition of the Sudan in 1929*. London: His Majesty's Stationery Office.

Sudan Government. 1955. *Report on the Administration of the Sudan in 1951/52. Furnished by the Governor-General to Her Britannic Majesty's Government in the United Kingdom and to the Egyptian Republican Government*. Khartoum: Sudan Government.

SWECO International. 2003. *Review of the Implementation Status of the Trans-African Highways and the Missing Links*, vol. 2. Stockholm: The African Development Bank.

Verhoeven, Harry. 2011. '"Dams are Development": China, the Al-Inghaz Regime & the Political Economy of the Sudanese Nile.' In *Sudan Looks East: China, India and the Politics of Asian Alternatives*, edited by Daniel Large and Luke Patey, 120–138. London: James Currey.

Weber, Eugen. 1976. *Peasants into Frenchmen: The Modernization of Rural France, 1870–1914*. Stanford: Stanford University Press.

West, Louis C. 1918. 'Dongola Province of the Anglo-Egyptian Sudan.' *Geographical Review* 5 (1): 22–37.

Woodward, P.R. 1991. 'Government and infrastructure.' In *The Agriculture of the Sudan*, edited by G.M. Craig, 138–147. Oxford, New York: Oxford University Press.

WHO. 2009. *Global Status Report on Road Safety*. Geneva: World Health Organisation.

Index

www.ingramcontent.com/pod-product-compliance
Lightning Source LLC
Chambersburg PA
CBHW071431230926
51846CB00001B/183
9789004336742